STUART STRUEVER, director of the Koster excavations, is former chairman of the Department of Anthropology and Archaeology at Northwestern University and former President of the Society of American Archaeology. He is considered one of the foremost exponents of the "new" archaeology.

FELICIA ANTONELLI HOLTON, a free-lance journalist, is a former reporter for the *Wall Street Journal*, and a national prize-winning editor of *The University of Chicago Magazine*.

MENTOR Books of Special Interest

KOSTER

AMERICANS IN SEARCH OF THEIR PREHISTORIC PAST

STUART STRUEVER AND
FELICIA ANTONELLI HOLTON

A SIGNET BOOK

NEW AMERICAN LIBRARY

TIMES MIRROR

To
Alec Helton,
Mary and Teed Koster,
all our friends and neighbors
in Greene and Calhoun counties, Illinois,
and to their long-ago predecessors,
the Koster Amerindians,
this book is dedicated with affection.

Copyright © 1979 by Stuart Struever and Felicia Antonelli Holton

All rights reserved. For information address
Doubleday and Company, Inc., 245 Park Avenue,
New York, New York 10017

Library of Congress Catalog Card Number: 80-80066

Painting by Jay Matternes from *The World Book Year Book*.
© 1973 Field Enterprises Educational Corporation

This is an authorized reprint of a hardcover edition
published by Doubleday and Company, Inc.

SIGNET, SIGNET CLASSICS, MENTOR, PLUME, MERIDIAN AND NAL
BOOKS are published by The New American Library, Inc.,
1633 Broadway, New York, New York 10019

First Signet Printing, May, 1980

1   2   3   4   5   6   7   8   9

PRINTED IN THE UNITED STATES OF AMERICA

Contents

I

The "Arkies"

1

The Secret in the Cornfield

It looked just like any other cornfield.

It was, in fact, just an ordinary cornfield on a farm in west central Illinois, one among thousands that dot the plains of the American Middle West. The cornstalks, green and shimmering in the late-summer sun, were about ten feet high. They covered three acres of a gentle slope which ran from the base of 150-foot-high bluffs down toward the white frame farmhouse. The bluffs were heavily wooded, but here and there among the dense greenery a patch of limestone reflected the sun's intense light.

Unknown to Alec Helton and me as we stood at its edge on a hot August afternoon in 1968, the cornfield held a secret.

It had kept its secret so well that it had remained hidden for more than nine thousand years.

As we stood there, the two of us had different expectations.

Alec, a farmer, who is an enthusiastic collector of Amerindian artifacts (archaeologists call anything made by human beings an artifact), thought the fields of his next-door neighbor, Theodore Koster, might contain an important archaeological site.

I am an archaeologist. I was there to please my friend Alec. For some time he had been nagging me to examine Koster's fields. I figured we would spend half an hour there and would find the same kind of artifacts on the surface of the field that could be found on a dozen fields throughout the Lower Illinois River Valley. I would then put the Koster farm on a long list of potential archaeological sites which, someday, I might excavate. Then, I hoped I could go home, have a cold beer, and enjoy Alec's friendship without being nagged.

Alec (who hasn't been called by his given name, Harlin, since his christening day sixty-four years ago) wore faded blue denim overalls and a blue work shirt. On his head he wore the identifying badge of a west central Illinois farmer, a billed orange cotton cap, with an embroidered patch on the front of the crown which read "Golden Harvest," the brand name of a hybrid corn. As we stood there, Alec lifted his cap to scratch his head, revealing a stark white bald head, fringed with graying light brown hair. The white skin on his head was in sharp contrast to his deeply tanned face and hands.

It gets very hot in the Lower Illinois River Valley—or, as my colleagues and I call it, "Lowilva"—and I like to wear khaki cotton pants and shirt to work in. Frequently I'm taken for a farmer, which suits me fine. I grew up in the small town of Peru, Illinois, and spent my boyhood in the fields outside of town, so I feel very much at home with rural people. At the time, I was thirty-nine, an associate professor of anthropology and archaeology at Northwestern University in Evanston, Illinois.

I had been spending my summers since 1958 excavating sites in Lowilva. My main research interest at the time was a group of prehistoric cultures called Hopewell, which had lived in Lowilva between 100 B.C. and A.D. 450.

I had met Alec in 1962, when I was digging a Hopewell site at Apple Creek, about ten miles north of the Helton farm, in Greene County, Illinois.

In 1962 the state of Illinois had been working on the Hillview-Eldred road, which runs north-south on the east side of the Illinois River. The workmen were putting blacktop on the gravel road and, in the process, straightening the road. The new road cut right through the Koster farm and down through the next farm, which is Helton's. As the bulldozer scraped earth from the farmlands to clear a roadbed, just about two miles south of the Koster farm it uncovered concentrated masses of fire-reddened rock, mussel shells, bits of animal bones, occasional projectile points (stone points used for spears), and potsherds (bits of broken pottery). To the average eye, these items, encased in dirt, looked like bits of rubble, indistinguishable from the rest of the debris in the soil. But to the experienced eye of a professional archaeologist, or of a skilled amateur archaeologist or artifact-collector, these signs indicated that prehistoric people had once lived there.

When I heard that the state was running a bulldozer through the two farms, I sent a student to look at the road cut. The student picked out what might have been several archaeological sites along the cut. Unfortunately, these were now destroyed forever.

Someone else also looked through this material and recognized it for what it was—Alec. At the time, we had not met, but he was aware that I was working in Lowilva, and he decided to come and tell me about the sites.

When Alec told me about his observations, I explained that we were too late to save the archaeological sites which the road-builders had cut through. I suggested, however, that if he came across any more potential sites, he tell me about them before they could be destroyed.

In reply Alec suggested that I take a look at the Koster farm. Back in 1943 Alec's farm had been flooded when a levee on the Illinois River had broken. He had moved his family temporarily into the tenant's house on the Koster farm. As the flood went down, Alec noticed artifacts lying about on the ground near the tenant's house, where they had been exposed when the receding waters carried away topsoil.

Over the years since then, Alec had collected many more artifacts from Koster's farm, with his neighbor's permission. He had been a collector since boyhood; he goes artifact-collecting every spring after the rains. When his two daughters were young, they accompanied him. Now his grandchildren go on the annual spring artifact hunt with him.

There is a big spring behind the Koster farmhouse, out of which flows a creek. The west bank of the creek is heavily eroded, and it was there that Alec kept finding artifacts.

"You really ought to see this place," he told me. "There are artifacts all over the place. On that hillside where Teed [Theodore Koster's nickname since boyhood] grows corn, there are lots of things. The rains wash a gulley about six inches to a foot deep every year down each of those corn rows and expose stuff. There are some things in the hog lot too, because the hogs root out any vegetation and the ground erodes."

After that, Alec began to drop by to visit me at the Apple Creek site and we became friends. Alec helped us in many ways. Like many farmers, he is a self-taught auto mechanic. If a truck broke down, he volunteered to fix it. Once, when we needed an outhouse, Alec remembered having seen an available one, and fetched it in his pickup truck. To the delight of the students, he constantly brought us gifts of fresh-picked vegetables from his own garden.

In exchange, I lent Alec books on archaeology and took time out on his visits to explain the work going on at the dig. Alec was eager to learn how to identify the artifacts he found with the cultures that had produced them. When he decided to set up a little museum in his basement, I sent a student to catalogue his collection by archaeological culture and site.

One consequence was that Alec began to realize that there was a wide diversity in the ages of the projectile points and other artifacts he was picking up on the Koster farm. He told me about it but I didn't react the way he expected me to, partly because I was a bit skeptical that he was finding objects of such diverse ages in one place, and partly because I didn't think this necessarily meant that people had lived there very far back. Projectile points have been found on the prairies of the Midwest, but no habitation sites have been found there. The spear points appear to have been lost at random, possibly by hunters passing through.

Besides, Alec was only one of about a dozen people who were stopping off from time to time to show me artifacts they were finding in the area. These local collectors have been extremely valuable in helping us locate many sites.

Over the years I have developed a warm friendship with many of these people, whom I've come to call "archaeologists on the landscape." During my first years in Lowilva I used to call on them twice a year, on a regular basis, to record their finds. In time my visits to their homes became an informal ritual. I would drop by, and we would sit around the kitchen table and chat. Frequently I would be invited to share a meal—deep-fried catfish or buffalo fish, home-grown vegetables, and homemade pies. Then the collector would haul out a bag of artifacts. I would record the size and shape of each item on a data sheet, with information about where it had been found. These finds give clues to which people had lived at different sites in prehistoric times. On my return to the university, I would file the data forms under "potential sites."

Today there are about twenty-five "archaeologists on the landscape" who help the Northwestern Archaeological Program (NAP) locate sites. They use NAP recording forms, artifact bags, and artifact tags just as we do. Staff archaeologists make the rounds for me, visiting the collectors. Some of the artifacts are donated to NAP, but if a private collector wishes to keep his finds, the archaeologist makes a photo or drawing of them.

Of all the people who were stopping by with information about potential sites, Alec was by far the most persistent.

Part of my resistance to Alec's oft-repeated request that I look at the Koster farm stemmed from my own frustrations. I was as interested in potential sites as Alec and the other collectors, but I simply did not have the resources to explore them. The best I could do at that time was to list them in a file and dream about them as future work.

It took all of my time and energy to teach, lead a handful of students in digging and analyzing the findings from the sites we had discovered in Lowilva, and raise enough funds to keep the little group housed and fed each summer.

At most American universities archaeology receives a relatively small share of the total research dollars available. I was an unknown archaeologist, digging in an area of the world considered archaeologically uninteresting by most people. I had found that when I talked to people about the potential for excavation in Africa, their faces would light up. But when I said that I was excavating sites in Illinois: their eyes would glaze over in boredom.

Partly, this reaction stems from the fact that Euro-Americans have long been prejudiced toward their predecessors, the Amerindians. From the first encounters of white Europeans with American aborigines down to the present day, many people have assumed that because the natives had a technology that was simpler than theirs, the Amerindians were simple-minded, and consequently not much could be learned from studying their past.

One day in the late summer of 1968 Alec's son-in-law, Joe Brannan, who farms Koster's land for him, and Joe's son, Stanley, found two drilled "plummets" lying on the surface of the ground, right in the middle of the Koster north cornfield. These objects have puzzled archaeologists for a long time. They are teardrop-shaped stones, polished to a smooth finish. The ones the Brannans found were about three inches long and one inch across at the widest point. Each had a neat hole drilled through the narrow end, large enough for a sinew to go through it. The Brannans gave them to Alec for his collection, and he brought them to me.

At this time archaeologist Frank Rackerby and I were excavating the Macoupin site—a Hopewell village site in the Illinois River Valley floodplain, about four miles south of the Koster and Helton farms.

When I saw the plummets, I was immediately interested. I called all the students around to examine them, and explained their importance to archaeologists. Similar objects have been found in several places in North America, and archaeologists have attempted to explain their possible use in many ways. Some think they were used as bolas are in South America—tied to the ends of thongs and thrown around the legs of a running animal to capture it. Archaeologists excavating in California have called them "charm" stones and hypothesized that aborigines wore them tied to clothing or draped on sinews about their bodies.

I think artifacts of this type were used as weights on the ends of cast nets, which were thrown over waterfowl or other birds in hunting; and that is why in NAP we refer to them as plummets.

The plummets found in Lowilva are beautiful forms made of imported iron ore, either hematite or magnatite. The closest source of hematite to the valley is at least fifty miles away, in southeastern Missouri, near present-day iron mines.

After talking to the students about Alec's latest acquisitions, I was suddenly very curious about the Koster farm. I told Alec I'd go look at it with him when we closed down for the day. About four-thirty that afternoon I quit work and we drove to the Koster farm. The farm is located in Greene County, Illinois, about fifty miles northwest of St. Louis, Missouri, and about 270 miles southwest of Chicago.

At that time, in late August, the corn towered several feet above our heads. We stepped in between the rows, and immediately found ourselves in a hot, dark place. The temperature that day was 100 degrees F.; among the corn-stalks it must have been about 110 degrees. Within sec-onds we both were sweating profusely.

As we began to move through the corn, keeping our eyes on the dry brown earth, I put up my right arm to protect my face from the nagging pricks of corn leaves. Their edges are razor sharp, and some were right at eye-ball height. I was wearing a short-sleeved shirt, and my arm soon was covered with red welts where the corn leaves stabbed it as I held it up to shield my eyes.

In addition, we were moving through clouds of pollen dust. Corn pollen usually is not blown away; it falls directly down on the plant. Most people are not allergic to corn pollen, but they do react to quantities of it. I felt as if a ton of pollen had been dropped on us, and I began to sneeze. As we worked our way through the stalks, my arms itched, my skin felt as if it were on fire, and my eyes watered from the pollen.

"Nobody in his right mind would be doing this on a day like today," I muttered. "It's too damn dark in here to see anything."

"Dammit, Stuart, you have no patience," retorted Alec.

But almost immediately we began to pick up quantities of Jersey Bluff pottery, and I forgot my miseries in the ex-citement of the search. I knew that the Jersey Bluff people, who lived around A.D. 800–1200, had been recorded at sites in the area, and it was obvious that they had lived here too. The broken pieces of pottery which we were picking up generally varied from the size of a penny to that of a silver dollar. Once in a while we would pick up one that was twice the size of a silver dollar. I paused for a moment to wipe the sweat from my face and to examine several jagged pieces of pottery before placing them in a bag.

Over the years I had become an expert on prehistoric ceramics, and I can identify a wide range of pottery from the American midwest. Jersey Bluff pottery is distinctive; it is orange, thin-walled, and of a uniform thickness. The vessels vary a lot in size and have slightly sloping shoul-ders and widely flanging rims. The pottery has black tem-pering, consisting of hornblende, a crystalline rock found in gravels brought to the Midwest by glaciers. The Jersey Bluff people added hornblende to clay to give some firmness to the vessel's walls while allowing water to evaporate and escape from the clay without exploding the pot when it was fired.

Among the artifacts we were picking up were tiny arrowheads made of white flint, each about one inch long. These were characteristic of the Jersey Bluff people.

Most people, when they think of Amerindians, assume they always used bows and arrows. Actually, during most of prehistory in North America the aborigines used spears, tipped with stone projectile points. It was not until about A.D. 800 that they began using bows and arrows.

All archaeologists would accept the conclusion that people came to North America at least fifteen thousand years ago. New evidence from Alaska, Mexico, and South America points to a much earlier arrival in the New World, probably at least thirty thousand years ago, and maybe before. During the long time span from the first arrival of people in North America until the early 1500s, when the Spaniards appeared and introduced iron in what is now the American Southwest, the Amerindians depended largely on stone and bone for their weapons. And up to A.D. 800 stone projectile points were produced in a limited range of sizes, although the styles varied considerably. Most projectile points were from two to six inches long, and were used as points for spears. The spears were thrown from the hand, or from *atlatls* (spear-throwers) made of wood or animal bone.

All of a sudden, about A.D. 800, smaller stone points began to appear. Archaeologists believe that the change in size of points corresponds to a shift from the use of spears to bows and arrows. A large stone projectile point is too heavy to be used on an arrow very effectively. Many small white arrowheads made of Burlington chert, a form of quartz which can be found on the Koster farm, were lying around in the cornfield.

We also came across pieces of discoidals, which probably had been broken up by the plow. A discoidal is a biconcave wheel-like object made of granite or diorite. These beautifully polished pieces of stone—ranging from two to four inches in diameter—were used for the game of *chunky*, which early English and French explorers observed being played by Amerindians in the American Southeast. In the game, the chunky stone was rolled on edge and the players threw spears at it to score a hit. The

Jersey Bluff people were the first makers of chunky stones in the area.

We continued our search, going over the cornfield south of the Koster house, the hog lot, and finally, the rest of the farm. We didn't stop until it became too dark to see. Exhausted and sweating, we headed for the tap in Teed's back yard and refreshed ourselves with cool spring water.

I figured that the Jersey Bluff village which once had flourished where the Koster farm now stood must have extended over twenty-five or thirty acres. The largest Hopewell village site I had ever seen covered about six or seven acres.

I was intrigued. I had spent several years intensively studying the various Hopewell cultures. Here on the Koster farm was apparently an excellent example of one of the cultures which came after Hopewell. Jersey Bluff was a culture which had been able to organize its population into much larger communities than Hopewell. The Jersey Bluff people were the first in Lowilva to develop large towns, such as the one we had just found, and were also the first to establish major occupations in the tributary valleys leading out of the main valley of the Illinois River.

Archaeologists speculate that the population in Lowilva was growing sharply at about A.D. 800 and that some people were being forced to move into marginal areas, such as the secondary valleys, which contained fewer natural food resources than the main valley.

Now, here I was, standing next to a village site which was larger by far than those of earlier cultures in Lowilva. Perhaps this site would give us a chance to test some theories about the growth of population in the valley.

Besides, among the artifacts Alec had shown me which he had found on the Koster farm I had recognized some from prehistoric cultures that had lived in Lowilva before the Jersey Bluff people.

"Alec, you were right," I said. "We've got to dig this place, to study Jersey Bluff, and to see if there's anything older underneath."

Alec, who had been leaning over the faucet, pouring cold spring water over his head, looked up at me and, with rivulets of water cascading down his tanned face, gave me a great big smile.

2

"Some Day You Ought to Dig There"

That winter, back at my duties on the Northwestern University campus, I began to recall my friend Gregory Perino's earlier observations about the Koster cornfield.

Greg, then staff archaeologist for the Thomas Gilcrease Institute of American History and Art in Tulsa, Oklahoma, is a self-trained archaeologist who has gained the respect of other professionals in the field. He began hunting for artifacts during his boyhood in Belleville, Illinois, after reading James Fenimore Cooper's Leatherstocking Tales. While in high school, he would cut classes and hitch rides on freight trains over to the Mississippi River Valley in Illinois, where he could find plenty of artifacts. Until he was forty, he worked full-time as a machinist at Scott Air Force Base in Belleville, Illinois. He spent weekends and vacations excavating, and over the years became an expert. He specializes in mortuary sites, and has probably dug more prehistoric cemeteries than anyone else in the history of North American archaeology.

In 1952 the late Thomas Gilcrease, a pioneer oilman from Tulsa, who was part Creek Indian, approached Greg and asked if he would like to work for the Gilcrease Institute. For the last twenty-six years Greg has been a full-time archaeologist, digging in Arkansas, Illinois, Missouri, and Oklahoma.

A mild-mannered, soft-spoken man who carries the stamp of his Italian forebears on his olive-skinned features, Greg is a spellbinder when he begins talking, in a soft voice, about his work. He is a walking encyclopedia on the cultural habits of various prehistoric peoples who dwelled in several midwestern and south central states, as well as on the terrain and weather patterns of those states. He is also an expert on projectile-point styles, and has catalogued the collections at the Gilcrease Institute, and at the Museum of the Red River in Idabel, Oklahoma, where he is now curator.

Greg's appearance is deceptive. He is tall, thin, even a bit frail-looking, but he is in excellent physical condition from years of walking over all kinds of terrain and handling a shovel at excavations. He is twenty years older than I am, yet I've found myself running to keep up with him when he was enthusiastically in pursuit of some archaeological discovery. Because he was trained on the landscape instead of in the classroom, he has developed an extremely keen ability to "read" an area for hidden archaeological sites. Repeatedly, I have seen Greg take clues which would seem extremely scant to most archaeologists and process them mentally to come up with very sound conclusions.

Several years earlier, in 1962, when I was digging the Apple Creek site, Greg was excavating the Koster Mound Group, a series of seven prehistoric burial mounds on the bluff crest above the east field on the Koster farm. One day I came down to visit Greg's site. (Archaeologists love to visit each other's sites.)

I met Teed Koster for the first time that day, then climbed to the top of the bluff to visit Greg. Eventually Greg would discover about three hundred skeletons in the Koster mounds; on the day that I visited, he had just uncovered about a dozen skeletons, and we spent some time looking these over. Many years later, when the Koster site was dug, we realized Greg had been excavating the cemetery established by the Jersey Bluff people, who had occupied the top horizon at the site. ("Horizon" is the term for an identifiable layer or level in a site which was the location of a single prehistoric culture.) We walked to the edge of the bluff, and looked out over the valley, a view we both find beautiful.

Greg pointed to the Kosters' north cornfield and said, "There's a spring-fed creek down there, Stuart. I'm going to dig a hole by that spring, and I think there ought to be some deeply buried materials there, very early objects.

That field looks to me like the kind of place that would have been occupied during the Early Archaic period [8000–5000 B.C.]."

Over the years, Greg and I had tacitly divided Lowilva between us when we searched for archaeological sites. Greg dug burial sites; my interest is in habitation sites. But occasionally Greg would sink a test pit into what he suspected was a habitation site for clues to date material from the burial mounds he was digging nearby. During the Middle Woodland period, 100 B.C.–A.D. 450 (the time in which the Hopewell cultures had flourished), and the Late Woodland period, A.D. 450–1200, residents of Lowilva frequently buried their dead in mounds on the bluff crests not far from their villages below.

That fall, Greg wrote to me: "I found some Godar [projectile] points, several feet below the ground surface near the spring on the Koster farm. Looks as if there is a very early village there. Some day you ought to dig there."

At the time he wrote, Greg didn't know what the radiocarbon dates for Godar projectile points were, he only knew that these points came from a period much earlier than either Hopewell or Jersey Bluff.

I filed Greg's letter and had forgotten all about it until the winter following my first exploration of the Koster fields.

I spent most of the next summer (1969) working in the laboratories at Northwestern University, analyzing material from the Macoupin site, which we had dug the previous year. For every hour an archaeologist spends in the field, he or she spends two to five hours in the laboratory, trying to make sense out of what has been retrieved from the site.

One day toward the end of August I walked into the laboratory and startled my students by declaring, "We're going digging tomorrow. Somebody call Vern and see if he wants to join us."

I had become increasingly preoccupied with my observations of the Koster cornfield from that hot day the previous summer, and with Greg's earlier remarks, which I now recalled vividly. All summer, I had disciplined myself to the task at hand, the analysis of the Macoupin site data. Suddenly, as the summer drew to a close, I realized I was very curious to see what the Koster fields might contain.

"We" consisted of myself and anybody in and around

the laboratory who felt like putting in three weeks of hard labor at no pay. That was all the time left until the fall quarter began at Northwestern University. Several students promptly offered to forgo their vacation period in order to dig. Doris Evans, a Northwestern University secretary, begged for a chance to exchange her typewriter for a shovel, and she brought a friend, Doris Bergman, a Des Plaines, Illinois, homemaker.

"Vern" was Vernon Carpenter, now seventy-eight, a retired operator of a tire-recapping plant, from Posey, Illinois. Several years ago Vern became crippled by arthritis and had to use canes, and sometimes crutches, to walk; he became very depressed. He is an artifact-collector, and at a meeting of collectors he met Greg. They became friends, and Vern managed to hobble out to visit Greg in 1961, while he was digging the Klunk mounds near Kampsville, Illinois. Fascinated, Vern asked if he could help. Greg set him to screening dirt for artifacts. He left his canes beside a tree and began to work. Within the next few days, Vern found he could walk without his canes, and he hasn't used them since.

That summer Vern eagerly accepted the invitation and met us in Kampsville—about nine miles from the Koster farm—where we set up living quarters for our three-week stay.

Vern shared a room with the young male students. He would wake up in the middle of the night with horrible leg cramps, and his roommates learned that the best thing to do was to jump out of bed and knead his leg muscles very hard. That got the circulation going and relieved the pain. When I heard about this from Vern, I marveled at his charm in winning acceptance from the young men, and at the students' kindness in helping their fellow worker.

Teed Koster had given me permission to dig some test pits. Ordinarily, before doing any digging at all, we conduct a very controlled surface survey of an area we are considering for excavation. In a rural area this is made easier because farmers have plowed the surface of many ancient sites in North America, and exposed the remains of prehistoric communities. Since every human community is structured by the activities that take place in it, the location of debris on the surface reflects where certain activities occurred within the community. When we survey a surface, we lay a grid on it, using string and small stakes

to mark it out. Next we go over the ground surface, collect the artifacts from each square in the grid, and bag
them separately. Then we go back to the laboratory and
plot these findings on graph paper, square by square, to
see where there were differences in human activities on the
site, as reflected in the kinds and amounts of artifacts
found on the surface. From that information, we decide
where to dig test squares. The ultimate goal is to sink
squares in each area where the surface debris indicates a
different activity took place.

Usually it takes five or six weeks to lay the grid, search
for artifacts, wash them, analyze them, and plot them on
graph paper. Since we had only three weeks and a limited
amount of money, I decided to skip this step and rely on
my observations from the previous summer, when Alec
and I had done a surface survey. I remembered distinctly
where we had found concentrations of flint chips, potsherds, and other artifacts. After a careful survey of the
surface, I picked out three locations in which to sink test
squares.

From my earlier examination of the farm I knew that
the Jersey Bluff village had covered a great deal of the
cornfield to the south of the house and the pasture sloping
up to the east of the house. We also had found some Mississippian artifacts there, from a period (A.D. 900–1673)
overlapping Jersey Bluff. Accordingly, I decided to sink
two test pits in each of these places, since I was curious
about the cultures that had come after the Hopewell in
Lowilva.

I carefully examined the patterns of erosion on the Koster farm. In eroded areas, if there is debris left from a prehistoric site, it may have been washed away from the
original spot in which it was dropped, which then distorts
the picture. The rate of erosion varies from place to place,
depending on the slope of the ground and whether or not
the farmer has set up anything to block the erosion. If the
farmer has planted a row of bushes on the slope, he may
have altered the rate of erosion considerably.

I walked along the bank of the spring-fed creek, just
northwest of the Koster farmhouse, thinking about Greg's
suggestion that there might be material from earlier cultures buried there.

The north field (the cornfield which I had first explored
with Alec) was very eroded, partly because Teed had been

plowing there and partly because rain had repeatedly washed soil down from the bluffs. When we had surveyed the cornfield, we had found the density of prehistoric material much lower in this field than in the other two locations.

Yet, despite the evidence of erosion and the lower density of artifacts in the north cornfield, all of my instincts told me to sink some test pits there. As I stood deliberating, I bent and picked up some soil and sifted it through my fingers. Then I looked from the cornfield to the top of the bluffs, which stood several hundred yards to the north.

The soil I was fingering was *loess* (pronounced "luss"), a fine, dry powdery soil which had been carried from the west by strong winds associated with the glacial front during the last Ice Age. It had been deposited on the tops of the bluffs which line the Illinois River Valley, where it had built up in deep mantles. Over the years, the loess had been redeposited in the valley (and in the Koster cornfield) by rainwater, which washed it down the slopes of the bluffs.

I looked from the grains of loess in my fingers to the bluff tops overlooking the cornfield. Just possibly, I thought, rains had washed down enough loess to cover any debris that may have been left by prehistoric people. I decided to dig some additional test squares in the north cornfield to test this idea.

We began to work, first marking out several isolated squares at different points among the tall corn in the north and south fields, and in the pasture on the east hillside. Archaeologists dig in standard units to enable them to keep careful records. We dig in squares six feet on a side, and in three-inch levels. This is an arbitrary choice; at various excavations archaeologists may dig in different-sized squares, and in different depth levels. What is important is to dig in equal, measured levels, and to record carefully what is taken from each level since this information will be used later for data analysis. If the excavator encounters a definable change in the soil corresponding to a cultural feature (a nonportable artifact, such as a hearth, house floor, or pit), he or she is then guided in recording and removing material by the beginning and end of that cultural feature, not by an arbitrary three-inch level.

The August heat was stifling; there seemed to be no air at all in the standing corn as we shoveled. It was like being put into an earthen box from which there was no escape and having someone turn on a hot lamp over one's head. And after spending the summer in the laboratories, we were all pale-skinned, and our muscles were soft.

As we dig, we toss dirt out of the hole onto screening tables, which are rectangular wooden tables covered with one-half-inch wire mesh. (If the dirt at a site is very dry, we may switch to one-quarter-inch mesh.) As the dirt falls through the screen, everything larger than one-half inch in size is caught on the screen. Screeners standing next to the tables remove these items and place them in plastic bags. The bags are tagged, to show what square and what level of that square the item came from. Some of the dirt that falls through the screens is caught in half-bushel baskets set below the tables. These baskets are similarly tagged. At the end of the day the bags of artifacts and baskets of dirt are taken to laboratories for further processing.

First we took out the plow zone, which is the first nine or ten inches below the surface, soil that has been turned over by the farmer. If there are artifacts in the plow zone, they may have been broken into pieces by farm machinery.

After we had disposed of soil from the plow zone, we went through about eight or ten inches of organically stained midden deposits. Organic staining results from the decomposition of food remains or from feces, and it usually indicates that human beings have lived at the spot. "Midden" refers to the layer of refuse found at a prehistoric site. In this midden we found burnt limestone, animal bones, mussel shells, and chipping debris of the same Burlington chert from which the arrowheads we had found had been made. Chipping debris, of flint or chert (which are quite similar forms of quartz), is always a sign that tools have been manufactured from these materials at the spot. The debris we were finding was similar to the Jersey Bluff material Alec and I had found on the surface a year earlier.

At the bottom of this midden we came to sterile base soil. This was loess and contained no cultural debris. It was striking in comparison to the layer above because it was a light buff color, homogeneous in texture, with no stones or other materials in it.

When we examined the sterile loess, it looked quite undisturbed by human beings, so it would appear that we had reached the bottom of the site. But, as I have mentioned, in rapid slope-wash conditions, when rainwater comes down the bluff it frequently carries soil with it which it redeposits at the bottom. Therefore soil in such deposits may only look like a base soil that has been there for a very long time. I decided to use a probe.

The probe that we use was developed by Greg. It consists of a spring-steel rod about three feet long attached to a shovel handle. At the tip of the rod is a ball bearing. The prober pushes the steel rod into the ground as far as it can go and rotates the ball bearing so that it picks up a bit of soil for us to examine for its color and texture.

As I pulled up the first probe I saw that my hunch had been correct. Instead of the buff-colored earth on which we were standing, this soil was black. Apparently there was a cultural layer below. We kept on probing.

Over and over again, down went the probe. Each time, we moved it a few feet, sampling the ground beneath the sterile layer. Suddenly the soil caught in the ball bearing showed tiny flecks of orange in it. Very carefully I removed the bit of soil from the ball bearing and spread it out on the palm of my hand.

As the students bent over my hand to examine the orange and black bits, I explained: "Those look as if they might be bits of burnt clay. And see, those look like bits of charcoal. it's beginning to look as if we may have hit the remains of someone's ancient fire. Let's keep digging."

But we had to go through another three feet of sterile base soil before we hit dark soil again. Sweating profusely, we kept working. When those first few shovels of dark earth were tossed up and over the screens, our enthusiasm shot up again.

The transition was sharp. We went from light, buff-colored homogeneous soil immediately into a dark-stained midden, with burnt rock and bits of flint chippage, pottery sherds, and artifacts in it.

We paused to scrape down a wall of test square eight, in the north cornfield, to "read" the profile. As archaeologists dig down, they carefully examine the sequences of strata in the soil to help decipher the story of what went on at a site. From this profile we could see that modern farming had dug up the soil for the top nine inches, turning over the earth and churning up the debris left by the Jersey Bluff people, who had lived there long enough to build up a foot or two of deposit. I designated that layer Horizon 1. Before the Jersey Bluff people had come to the site, while no one had occupied that spot in the valley, soil had washed down the slopes from the north and west and deposited three feet or so of sterile soil. But prior to that deposition of soil from the bluffs, the Black Sand culture, circa 200–100 B.C., had lived there, as we could tell from the artifacts we found. I designated that layer Horizon 2.

Horizon 2 turned out to be very thin, only a few inches thick, indicating that the Black Sand people had lived there for much less time than their successors, the Jersey Bluff people.

Beneath Horizon 2, we again ran into a layer of buff-colored sterile soil, several inches thick.

The next occupational layer, Horizon 3, turned out to be very diffuse. It did not have black organic staining and was only slightly darker than the buff-colored sterile soil.

Vern and Carl Udesen, a Northwestern University graduate student, were digging in one test pit together, and they got into a violent argument about whether or not there actually was a cultural deposit which could be labeled Horizon 3.

I was working in a nearby test square and could hear their argument, in spite of the thick stand of corn between us, so I went over to their square.

"What the hell is going on here?" I asked.

"He insists that this is a horizon, but it isn't," said Carl. "It's sterile soil."

"Dammit, it has artifacts in it!" Vern insisted.

"Well, you're into some other kind of occupation," I explained. "They just weren't doing things that resulted in a lot of organic waste, that's why you don't have decomposition in the soil. Horizon 3 probably represents a number of brief occupations. Organic staining results when people stay awhile in one place and their garbage accumulates. People who come and go don't leave much garbage."

Later we found my assessment had been correct. Analysis of artifacts from Horizon 3 showed that people from the Riverton culture had lived at Koster for a brief time in about 1500 B.C.

Unlike Horizons 1 and 2, Horizon 3 had no pottery sherds in it, which meant that the people who lived there were probably from Archaic times—before about 500 B.C., when people in Lowilva had learned how to make pottery.

This was a very exciting discovery.

Archaeologists have designated several time periods in North American prehistory: Pre-Paleo-Indian, anywhere from about 30,000 B.C. to 12,000 B.C.; Paleo-Indian, 12,-000–8000 B.C.; Archaic, 8000–500 B.C.; Woodland, 500 B.C.–A.D. 1200; Mississippian, A.D. 900–1673. (Woodland and Mississippian overlap). The Archaic period is further broken down into Early Archaic, 8000–5000 B.C.; Middle Archaic, 5000–2500 B.C.; Late Archaic, 2500–500 B.C.

Hence, when we found no pottery in Horizon 3, I was elated. I knew, without any radiocarbon dating of materials, that we were finding remains from the Archaic period.

Just how far back in the Archaic period the site went, we could not yet tell. To date the site, we would have to send bits of charcoal taken from each horizon to a special laboratory to be tested by the radiocarbon method. As usual, we had no funds for these tests and had to wait until the end of the 1970 digging season, a year later, to find out how long ago people had lived at Koster.

As the holes grew deeper, I examined them to see whether my choice of test-square locations had been accurate.

In the south field and up in the east field, on the hill, the diggers soon found that Horizon 1 was as deep as the site went. There did not seem to have been any occupations in those places before the Jersey Bluff people established their large village there. So I shifted strategy and opened up more test squares in the north field. In that brief three-week digging season we went through six cultural horizons there.

Maybe it was luck, maybe it was my long years of experience as an archaeologist, but when I chose a spot in which to sink one test square (number eight), my aim had been extraordinarily accurate. Test square eight, it turns out, goes right through all the major horizons at Koster, each of them (except Horizon 1, in which there is some intermingling of cultural material) separated very neatly from the one above and the one below by a layer of sterile soil. As one goes down, layers of buff-colored soil alternate very clearly with layers of dark brown, sometimes black, soil. The dark layers were created when people lived at Koster intermittently over the centuries and dropped debris as they went about their daily lives, just as we drop pop-can tabs or gum wrappers. Because we could dig for only three weeks in that late summer of 1969, we went only as far as Horizon 6, but we suspected there might be more cultural material buried below.

The evening before we were to return to the Evanston campus, I sat at the edge of test square eight, my legs dangling over the side and gazed at the smoothed north wall of the square, "reading" the profile and thinking that the dark and light banding of the horizons was a beautiful sight. I wondered how I could communicate its message to other Americans, and how they would feel about it. My sense of isolation about the knowledge that lay at my feet, waiting to be explored, was heightened by the thick walls of dark green cornstalks surrounding the square and shutting out the rest of the world.

What we had discovered in the Koster cornfield, I realized, was a very unusual archaeological phenomenon. Since there were no pottery fragments in any of the debris layers below horizon 2, we knew that the bottom horizons must have been occupied either during or before the Archaic period. A great many Archaic sites had been excavated in North America, but the information about this period of prehistory was limited, for several reasons.

For one thing, what we knew about the Archaic cultures was mainly from stone artifacts, with little information about the environmental adaptations of these peoples, because until recently that was how archaeologists studied ancient cultures.

In the past fifteen years there has been a great deal of discussion, some of it heated, among North American archaeologists, as a small group of us have attempted to bring about radical changes in theory and methods. One of our major points is that archaeologists must study the environment in which ancient people lived, along with what they had made, in order to understand their lives fully. At some point someone labeled us the "new" archaeologists to distinguish us from traditional archaeologists, and the name has stuck.

Another reason not much has been known about Early Archaic cultures is that many of the sites have been found in caves and rockshelters, and the data they contained was not easily recoverable or was contaminated by falling rocks. Sites of this kind are extremely difficult to excavate. There may be rocks, some as heavy as five thousand pounds, which have fallen from the cave or rockshelter roof, while people lived there or after. These rocks have compressed the layers of occupational debris and disturbed the relationships of artifacts and other evidence. Cultural materials from one human group might easily have become mixed in with those from another group.

By contrast, here at the Koster site, after each prehistoric group left, slope-wash conditions soon deposited a layer of sterile soil from the neighboring bluff tops over the remains of the abandoned village or camping site. In almost all of the Koster horizons, the layer of deposited sterile soil was thick enough to protect the cultural layer beneath it from further disturbance by later arrivals. Thus, when subsequent human groups came into the valley and settled in the same spot, they did not dig into the remains of the preceding occupants.

Such well-defined stratigraphy is rare in a North American archaeological site. At most sites, occupations followed so quickly upon previous ones and so little sterile soil was deposited between occupations that the cultural remains from one group usually mixed with those from another, so that archaeologists now find it difficult or impossible to separate them.

When I dug the Apple Creek site, ten miles north of Koster, in 1962, I could tell there had been five distinct occupations because I could identify artifacts from five different cultures. But because there was no sterile soil between the cultural layers, there was a great deal of mixing of artifacts from the different cultures, and so, beyond the artifacts themselves, there was not much opportunity to collect the kind of data we new archaeologists need in order to study prehistoric cultures.

A third aspect of Koster that made it special was the superb preservation of bone and plant remains in each cultural horizon. This was made possible because the loess at Koster is an alkaline (nonacidic) soil, which preserves animal and plant remains extremely well. These are vital clues to ancient peoples' environments.

At Koster, I could see, we would be able to do an intensive study of a long series of prehistoric village sites. The three factors of the stackup of many cultures over a long period of prehistory in one location, a well-defined stratigraphy, and the excellent preservation of animal and plant remains told me that we had found a unique and potentially important site.

Archaeologists digging in North America, unlike those who work in the Middle East or Central America, do not come upon great temples, gold caches, calendar systems engraved on stone, or any of the usual items that catch the public's attention. Instead, they try to decipher prehistory from dark-stained soils and small artifacts. They try to reconstruct a very detailed perception of the human chronicle in the New World, which is dramatic to tell but not very exciting visually. Hence, archaeologists in America are almost totally ignored, and in fact many Americans are unaware that there is much to be learned about the prehistory of their country.

In Koster I saw the opportunity I had longed for—a site that might appeal to the imagination of the American people. I felt that if other people could look at the profile at Koster and could be told why and how archaeologists excavate such places, they too would become as excited and moved about the exploration of American prehistory as I am.

3

The "Arkies"

Every year, toward the end of winter, I became very restless. I long for spring to arrive so that I can work outdoors again. I'm a "dirt" archaeologist. I like digging holes in the ground with a shovel. Many people have forgotten they have bodies and are unaware of the satisfaction of hard physical labor.

As the spring of 1970 approached, my restlessness was tinged with the suspense of excavating a new site. Long years of experience have taught me that one can never predict exactly what one will find.

I had spent part of the 1969–70 winter recruiting a larger staff to work in the laboratories I hoped to establish, and had expanded the size of the summer archaeological field school, since we would need more students to excavate Koster and to help in the laboratories.

In the late spring of 1970 we set up headquarters in the small country town of Kampsville, Illinois, located on the western bank of the Illinois River, a few miles north and across the river from the Koster farm. Since 1958, when I first began to dig in Lowilva, I had made many friends among the Kampsville residents. We also had used the town as headquarters each summer since 1964, renting various houses for our combined dormitory-laboratory-classroom setup.

As we moved in, forty-five strong, I could sense some apprehension among the townspeople. Our group was much larger than it had ever been before, and we were much more visible. Some of the houses we were using for dormitories were right on the town's main street, Illinois

Highway 100. The violent encounters between youth and police on Chicago streets during the 1968 Democratic convention were still vivid memories to a nation of television-watchers. Long hair on men was still a novelty, and several of our male students sported long hair and full beards. The custom of wearing one's oldest clothes to any social function had not yet become high fashion when our students appeared in Kampsville's streets in torn, faded jeans, some of the girls wearing "grannie" dresses, and people of both sexes wearing beads and going barefoot.

Kampsville's 450 citizens, descendants of the German-American Roman Catholics who founded the town in the 1840s, with a population heavily weighted with older people, viewed the youngsters with covert glances. Nonetheless, whatever their private misgivings, lifelong habits prevailed, and the townspeople maintained their custom of offering a polite "Good morning" or "Good evening" to everyone they met on the street, in a store, or a restaurant. The invaders, accustomed to the facelessness of their fellow citizens in American cities were startled, then charmed, with this local custom and eagerly returned the greetings.

Kampsville once was a major river port for the entire region. Steamboats plied the Illinois and Mississippi rivers loaded with its chief exports—lumber, hogs, and apples. The town is in Calhoun County, which occupies a narrow peninsula about forty-five miles long by six miles wide between the two rivers, which converge some thirty miles south of Kampsville.

The original settlers in the Kampsville area were Yankees straggling in from the east, followed by German immigrants in the 1840s, who settled in Calhoun County because its forests resembled those in their homeland. For several decades they made a living cutting and selling wood for steamboat fuel, for homes in St. Louis, and for barrel staves, railroad ties, and pulpwood. The land was valuable at the time. Now the forests have been depleted, and the steamboats replaced by trains and trucks. The tiny town of Kampsville has no industry to support it and has suffered economically.

Kampsville's lack of ecnomic and political clout is evident in the absence of a bridge spanning the Illinois River. Kampsville residents must use a small eight-car ferry to reach Greene County, where Koster is located.

The state of Illinois operates the ferry twenty-four hours a day, 365 days a year (weather permitting), free of charge. Once an urbanite has adapted to the pace of life in the valley, there is a certain piquant charm about this unhurried means of transportation. I must have crossed the Illinois River on the Kampsville ferry thousands of times by now, but I never tire of standing alongside the rail and looking upriver to see the same lush, verdant shores seen by prehistoric Amerindians, and later by French explorers as they paddled up the river in 1673.

Kampsville has only a few paved streets; graveled alleys make up the bulk of its thoroughfares. Capp's General Store (originally Kamp's, established by Captain M. A. Kamp, the town's founder), with its fifteen-foot-high walls and original embossed-tin ceiling, looks much the same as it did when it opened late in the nineteenth century, except for white porcelain and chrome frozen-food bins.

After we had been in Kampsville for a few weeks, that summer of 1970, I noticed that the townspeople began visibly to thaw toward the student crew, whom they had nicknamed "the arkies." Word had apparently spread, passed by those who crossed the river and pulled into Teed's back yard to take a look at what was happening in his cornfield, that these city youngsters had an awesome capacity for hard physical labor.

We lead a rigorous life during the excavating season. Each morning, six days a week, the Koster crew bus lumbers aboard the Kampsville ferry at 6 A.M.; it returns at 5 P.M. After a shower and dinner, the students attend lectures two or three nights a week as part of their academic requirements in the archaeological field school.

Archaeology is heavy physical labor, and in Lowilva it is done in extreme heat. The average temperature on a summer day ranges from 90 to 100 degrees F. People today have not always learned what their bodies can do, and one of the interesting by-products of an archaeological dig is the students' discovery of a physical capacity of which they have been unaware. In the early days of each season some newcomers want to quit, and it takes a watchful eye and some encouragement on the part of the supervisor to keep them going until they discover they can perform at a harder rate than ever before.

In 1970 I had managed to raise enough money to purchase a former hardware store and a seven-room house in

Kampsville. The rent for thirteen other structures, including houses, stores, and sheds, came to $550 a month (which is one of the reasons why Kampsville appealed to us as a headquarters town).

We set up three dormitories in old frame houses. None of the houses had running water. Toilet facilities consisted of wooden privies. I arranged for the crew to shower at Calhoun High School in Hardin, nine miles south of Kampsville. This meant the bus had to go nine miles out of its way, coming home from the site. It was a welcome relief, the following summer, when we could afford to pay the Kampsville school board for use of the showers at the local elementary school. The people of St. Anselm's Roman Catholic Church generously lent us the use of their church hall for a dining and lecture hall.

Tom Cook set up his artifacts laboratory in the town's former post office. And in an old building which had housed Bert Getz's hardware store for fifty-five years we set up a central processing laboratory for washing and cataloguing artifacts, bones, and other materials as they came from the field. Nearby was a seven-by-nine-foot shack which sagged wearily, its origins unknown. Manfred Jaehnig hung a carefully lettered sign, "Laboratory of Archaeo-Malacology," on one of its weather-beaten doors and moved in with the beginnings of his collection of prehistoric snails.

In the former drugstore on Highway 100 Fred Hill filled the display windows with what looked like handfuls of dirt spread out on newspapers. They were bits of animal bones set out in the window to dry after being washed. This was the first home of the zoology laboratory.

Every time I stepped across a worn wooden doorstep into one of these newly established laboratories, I was thrilled. They represented the beginnings of the implementation of an idea I'd had for a long time—a first step toward the establishment of what I hoped would become a permanent institute, located in a secluded place away from the home campus, to be devoted exclusively to experimentation in archaeological research and teaching.

We were already making a revolutionary break with tradition in archaeology by treating excavation and data analysis as a single unit, taking place concurrently. Ordinarily archaeologists ship excavated materials back to campus or museum laboratories, to analyze them during the winter season. If the data are to be analyzed by other scientists, the materials are sent to their laboratories; and then the archaeologist must wait patiently until the collaborating scientists find time to analyze them. It is not uncommon for materials to lie gathering dust on some scholar's shelves for years before someone gets around to them.

Our way, we could get feedback of information rapidly, to improve decision-making at the site while excavations were under way. One of the central goals of the new archaeology is to understand human behavior. For this, the archaeologists want to know specifically what activities took place in a village or hunting camp, and where. They place holes in the ground in locations where they suspect activity took place and collect samples of artifacts and "features" (nonportable artifacts such as fire hearths, house floors, tombs, or storage pits) that reflect those activities. Since the archaeologists never know exactly where the activities took place they may concentrate their diggings too much in one activity area and not enough in another. One way to correct that imbalance is to have feedback from the laboratories doing data analysis to tell whether the digging crew has obtained enough materials from one activity area and whether there are clues that they are beginning to get information from others. If the feedback comes quickly enough, the site supervisor can shift the crew.

At the beginning of that summer of 1970 I appointed two young graduate students, Kenneth Farnsworth and Gail Houart, to share the duties of supervisor for the Koster site. Ordinarily I serve as site supervisor, but it was apparent that my energies were going to be needed in several places that year, so I delegated the job to them, with the promise of frequent visits from me to see how the task was going and to help solve problems as they came up.

On the first day we arrived back at the Koster farm, I noticed Gail glancing over at the hogpen. She caught my eye, and we both laughed.

At the end of the previous year's three-week digging season we had not finished "taking down" Horizon 6 in test square eight (removing dirt to the bottom of the occupation level) and Gail had volunteered to remain behind and finish the job.

Early the next morning she had arrived at the cornfield alone, and was dumbstruck. After farmers in Illinois harvest their corn in the fall, they let hogs or cattle into the fields to feed on corn that has fallen from the machines. This field had been harvested the day before, and now, scattered among the knocked-down stalks, happily munching on remnants of corn, were about fifty hogs, several of which weighed more than three hundred pounds. Gail—five feet two inches tall, 110 pounds, and city-bred—was terrified.

"I'd never confronted a hog before without a fence between us," Gail later told me, "and I was scared. But I'd promised to do the work, so I started across the field. The hogs immediately left their munching to crowd around me, and, to my horror, they began eating my shoelaces. Now how do you get a three-hundred-pound hog to stop eating your shoelaces? I shook, but I kept going and finally made it to the snow fence we had put up around the open test pits to keep the hogs from falling in. By then they had eaten the bows from both sneakers."

Safe inside the snow fence, Gail discovered she had an even more vexing problem. The hole was six feet deep, and the only access to and from it was by means of a rope tied to a stake at the top. The stake was right up against the snow fence, and Gail realized with dismay that the hogs could put their snouts through the fence and nibble at the rope. It was apparent that they already had been chewing at it. She looked around but could find no safer place in which to position the stake. Feeling very low, she let herself down the rope. From the bottom of the hole she could hear the hogs grunting above and could see them trying to peer down at her through the fence. She was so anxious she kept making trips up and down the rope all day to check it, and she left the field early.

As she made her way gingerly through the snuffling hogs, she spotted Teed standing at the edge of the field.

"He had a big grin on his face, and it took me a few seconds to realize he was laughing at me," said Gail.

That night in Kampsville Gail met Dr. Jane Buikstra, associate professor of anthropology and archaeology at Northwestern University and director of NAP's biological anthropology program, who also had remained in Lowilva to do some work on the human bones she digs up from burial sites.

"If I don't come back to Kampsville some night, look for me in the hole," Gail told Jane. Then, having let someone else know that she might not make it out of test square eight, Gail relaxed and was able to work with more concentration for the rest of the week, though she did not manage to finish taking down Horizon 6 completely.

"I guess Teed would have rescued me if the hogs had eaten the rope, but he sure as hell would have laughed his head off while he did it," said Gail.

The first important decision we had to make about the Koster site in 1970 was where to dig in the cornfield. We reviewed the information from the 1969 test squares to determine where the major portion of the excavation should go. We decided to try to get the most information from Horizons 4 and 6. We knew from test square eight that these horizons were thick deposits, and they appeared to be *in situ* (in place) and undisturbed. We also knew some places *not* to dig for Horizons 4 and 6 since we had found a very low density of material for those horizons in some of the 1969 test squares.

We planned to continue digging deeper in the test squares already opened up and to dig more squares in the Koster cornfield to see what else we might find there.

The first surprise of the season came when Dana Omerod, a student excavator, mistakenly went beyond the bottom of Horizon 6 in square eight. He had been assigned to finish the job of taking down the square which Gail had not finished the year before. When a digger hits sterile soil, he or she is supposed to stop and consult the site supervisor. But Dana kept right on going, and after about three inches of sterile soil he hit another layer of debris. Puzzled, he called Ken, who came over to examine what he was encountering.

After consulting with Gail, Ken decided to test with a probe, this time using one with an auger at the end to bring up a small core of soil. He soon came up with evidence that there was another series of strata below Horizon 6 which bore evidence of occupational debris. I happened to come out to the site as they were examining the dirt retrieved in the probe. Elated, we counted the alternating bands of light and dark layers—Horizons 7, 8, 9, and 10. This told us, even from that scant evidence, that people apparently had come to this spot and lived there long enough to build up a debris in varying thicknesses on at least ten different occasions sometime in the past.

That evening in the dining hall there was a great deal of

excitement as the news of Dana's find spread to crew members who had been working at the laboratories in town or at other sites.

Around ten that night, after the evening's lecture for the field-school students, I met with Gail, Ken, Jim Brown—professor of anthropology at Northwestern University, who was codirector with me of the Koster Research Program and director of the computer laboratory for NAP—and some of the other research archaeologists over a beer at the Red Dog, a mangy little tavern in Kampsville, to discuss the implications of this new find for the program.

"Well, Stuart, it certainly doesn't look as if we're going to be able to dig this one in a summer," said Jim, grinning.

"Hell no," I replied. "At this point, I won't even predict how many seasons it's going to take us. But right now we're going to have to tackle some of the practical problems of dealing with a site which apparently is going to be very deep, and eventually very large in size. Some of our test squares are already thirty feet deep, and once we start opening up all the squares around them to get down to that tenth horizon, that's going to be quite a hole."

"How are we going to move all the dirt?" asked Ken. "It blows my mind to think about how many hours of back-breaking work it'll take to carry out that many bushel baskets of back dirt." (Back dirt is the soil left after diggers have screened the earth at a site. We put it to one side of the site until the excavating is finished, then push it back into the hole when we are through.)

"I've been thinking about that problem," I said, "and I'm going to ask Alec if he can help us find a solution for it."

We knew also that as we dug we would have to take precautions about the standing walls at the edge of the site. What would happen when we dug down and exposed tall walls of earth for long periods of time, as we planned to do? Would rain affect such walls and make them crumble? Most North American archaeological sites are relatively shallow, and we knew of no precedent to turn to for guidance.

In some of the deepest test squares, moreover, the diggers had been encountering mud as they dug Horizon 8. It was apparent that in some places we were hitting the water table (the level at which ground water settles). We would have to figure out how to de-water that portion of the site when we reached that level. One can dig mud, with a great deal of effort, but since it won't go through the screens, it's impossible to remove artifacts or other materials from it by that method.

And it was obvious that Koster was going to be a very expensive site to dig. For every foot deeper that you go below ground, costs go up, simply because it takes longer to dig deeper and to remove dirt from farther below ground level, and, in turn, it takes more money to house and feed the diggers.

Aside from these problems, there were other reasons Koster was going to be an expensive operation. We hoped to use the site to carry out experiments in the new archaeology. For their work, the new archaeologists collect massive quantities of material for analysis, much of which would be ignored by many traditional archaeologists. And, from the depth of the Koster site and the number of occupational levels to be excavated, it was apparent that the amounts of material to be retrieved would be enormous. We could see that it would take our scientists several years to analyze the data from so much material. We would need to house and support these scientists, supply them with laboratories and equipment, and provide storage space for the masses of material.

This plan was quite different from the way archaeologists traditionally work. Usually one archaeologist goes out to dig a site with a handful of students and operates on a very low budget. We planned to operate an archaeological center, with many co-operating scientists in residence each year for the excavation season at least. As we sat over our beer, we tried to figure out where we could obtain money to support our experiments.

"I guess I'll just have to hit the lecture circuit, to tell people about our work and try to raise funds that way," I concluded.

Early the next morning, Alec and I stood at the edge of the excavation, looking down at the diggers. I had asked him to visit the site with me, to see if we could come up with some ideas on how to solve our earth-moving problems.

We watched for a while. A student down in a hole would fill a bucket with dirt. Another student, at the top of the hole, would pull it up by rope, and dump it onto the screening table. Underneath the table sat the half-bushel baskets to catch the dirt that fell through. Students loaded some of these baskets onto a pickup truck to be taken to the flotation laboratory. They emptied the rest of the baskets into wheelbarrows, carted the dirt to the edge of the site, and dumped it.

"Alec, if we keep doing things this way, it will take us twenty years," I said. "There's no effective way to dig a hole this size without some sort of conveyor system. Do you think you could rig us up a conveyor for a couple of hundred dollars?"

"Well, Stuart, for a couple of hundred dollars you can't get much machinery, but I'll see what I can do," promised Alec.

We stood there for a while, watching the youngsters as they worked, and I felt myself relaxing a bit, for I knew that at least one of our problems was in very good hands.

4 ─────────────────────────

We Encounter an
Early Human and His
Best Friend

There were still more surprises awaiting us that summer. But, lest you get the idea that archaeological work consists of one spectacular discovery after another, let me assure you that, in between, there are hundreds of hours of back-breaking, tedious labor.

Take the experience of one of my students, Richard Rawlins. At the end of his first week of shoveling dirt at Koster he was excruciatingly bored. "For that entire week I'd been finding little but limestone," he recalled, laughing as he looked back on his experience. "I was disillusioned. I thought there ought to be more to archaeology than just endless shovels of dirt and limestone."

Rich had been chopping away with the spade at the thousands of lumps of limestone that characterize Horizon 6 at that point, and was getting tired of running into these obstructions. He happened to slice along the northeast corner of the square and exposed something that did not appear to be limestone. He stopped, wiped some dirt from the object, and thought it looked like a human fibula (leg bone). Since Rich had studied human and animal osteology, he thought he could tell the difference between human and animal bone.

"I dug around it with a trowel, and then a Perino pick,*

---

* A Perino pick is a bamboo pick used, because of its flexibility, to clear dirt away from delicate objects such as human bones. Greg whittles them and gives them to our students.

to get a good look at it," said Rich. " 'Yep,' I thought, 'it's a human fibula.' I didn't shout out, but I remember feeling quite happy about finding a bone in my square, as that is what I'd hoped to spend my summer excavating before I realized that one doesn't often find human bones on a habitation site."

Rich quietly informed Gail and Ken, who climbed down into the hole to take a look. They phoned me in Kampsville, and I drove over to examine this new find.

Meanwhile, news of the find had spread among the students digging at Koster.

Rich recalls: "Every member of the crew eventually popped in for a look at that marvelous fibula, all three or four inches of it. They had to come in one at a time, because we were using a knotted rope to get in and out of the squares, as you had not yet beguiled Sears, Roebuck into donating the aluminum ladders that we used in the following summers. I also had cut some foot holes into the wall of the square to use when I climbed in or out of the hole."

After I too had climbed in and out of the hole, Gail, Ken, and I agreed that we should consult Jane Buikstra, who was digging burial mounds a few miles away. She came by the next day.

After Jane had examined the bone, she informed us that it appeared to be part of an entire skeleton and the orientation of its axis suggested that the rest of the skeleton actually lay in an adjoining, unexcavated part of the site. At her suggestion, the bone was draped with wet newspaper, with a piece of canvas placed over it, to keep it from drying out. We agreed that in order to retrieve the rest of the skeleton, we would have to take down the three surrounding squares, and assigned Alice Berkson and Phillip Skager to work with Rich.

So Rich had to learn another painful lesson about archaeology. Even when you come across an exciting find, sometimes it takes more hard, dull work, time, and patience to unearth it. It took the combined efforts of the three of them six long weeks in the hot summer sun before they had removed enough earth from the adjoining squares to reach the rest of Rich's skeleton.

It turned out to be the complete skeleton of an adult

male. And Rich was given the satisfaction of removing those bones and placing them in boxes to be sent to Jane's human osteology (bone) laboratory in Kampsville for analysis.

At the time that Rich removed the skeleton, there were many questions about it that puzzled us. From our first look at the squares surrounding the skeleton, we thought there had been a village at that spot. So what was a human burial doing in a village? And why was there just a single burial? That it would take a great deal more digging and more analysis to answer these questions was the last lesson Rich had to learn that summer.

One morning at the site a girl's shrill scream rose above the thunking sound of shovels hitting solid earth. "It's a dog! It's a dog!"

A deeper, male voice joined her, repeating, exultantly, "It's a dog!"

Crew members dropped their shovels and trowels to crowd about the top of the square where the excited yells were coming from. Twenty-eight feet down, Nancy Wilmsen and Robert Kapka were jumping about in excitement at the bottom of the square they had been toiling in for five solid weeks.

Ken elbowed his way through the people on their knees at the edge of the square and swiftly dropped down to the bottom, using the knotted rope.

He knelt and examined the find. At the top, Gail had joined the group clustered about the hole.

"Yep, it looks like a dog all right," Ken shouted to the people above. "The skeleton's partly under a slab of clay, maybe a hearth. The bones look like they're all here, and he's lying on his side. Better send somebody for Fred."

Gail dispatched a student to Kampsville to fetch Fred Hill, the zoologist who had been assigned to analyze the Koster animal remains. Soon a green Honda roared up, and Fred jumped off and descended into the hole. By the time Fred emerged from the hole, I had arrived, and I too slid down into the square for a look.

"It's a dog all right, but a funny one," Fred explained, a few moments later, "because it's both big and small. It has a big head, almost as big as a German shepherd's, but the body is runty and the legs are strangely short. There's a good chance he was domesticated.

"Did you notice, there are two fire pits near the dog? He seems to be lying in one, but he's not burned, other than in the left front foot. It appears that he was placed on or near the fire pit after the fire had died down. His burial may have been of ritual significance."

The skeleton was well preserved except for the skull, which had been crushed by the weight of the overlying soil. The dog was between 45 and 50 centimeters (about 18 to 20 inches) tall at the shoulders, approximately the height of present-day fox terriers.

In fact, Fred thought that except for the head it strongly resembled the small Indian dog called the *techichi*, described by G. M. Allen in a 1920 study of the dogs of the Amerindians. Allen had described the techichi as a "small, light-limbed dog of rather slender proportions, narrow delicate head, fine muzzle. . . ." He also noted that this dog was described in many accounts of explorers as being "foxlike." From his study of the skeleton, Fred concluded that the dog was a mature adult.

Now we had another mystery to solve. What was the reason for the careful arrangement of the dog's body near the fire hearth?

The discovery of the ashes in the fire pit in which the dog lay was as important as finding the animal if not more so, for it provided a clue to when his human masters had laid him there to rest.

The question of how to obtain reasonably accurate dates for past events is vital to archaeologists. In the late 1940s Dr. Willard F. Libby, a chemist at the University of Chicago's Institute for Nuclear Studies, announced the development of a method to measure the amount of low-level radioactivity of carbon remaining in dead material of organic (living) origin. With this measurement it became possible to tell approximately when an animal or plant had died.

The carbon-14 (C^{14}) method of dating, as it is called, is based on the fact that radiocarbon has an atomic weight of 14 instead of the normal atomic weight of carbon, 12.

Radioactive carbon-14 is formed when powerful cosmic rays, traveling at high speed from outer space, bombard nitrogen atoms in the upper atmosphere high above the earth. Chemically, C^{14} seems to behave exactly like ordinary nonradioactive C^{12}. C^{14} atoms mix with oxygen in the earth's atmosphere to form carbon dioxide. This drifts down to earth, where it is absorbed by plants along with normal carbon dioxide containing C^{12}. Plants are eaten by animals and humans, who thus ingest the C^{14}.

Each C^{14} atom throws off a negatively charged electron and turns itself back into an atom of nitrogen. As long as the plant, animal, or human is alive, new C^{14} from the air keeps replacing the C^{14} changing to nitrogen. Thus, the amount of C^{14} in a plant, animal, or human remains about the same during life. (The amount of C^{14} in a person's body at any one time is infinitesimal.)

Once the plant, animal, or human is dead, however, no new C^{14} is taken in, and that which is already in the tissues keeps on breaking down and changing into nitrogen atoms. Like radium, uranium, and all other radioactive substances, C^{14} breaks down at a mathematically fixed rate. In scientific terms, it has a half-life of 5,730 years. This means that in 5,730 years, half the C^{14} in a dead plant, animal, or human will disappear by changing into nitrogen atoms. Half of the remaining C^{14} will disintegrate in the next 5,730 years, leaving only one quarter of the original amount.

Wood (if preserved, as it is occasionally at dry sites), soot, grasses, feces (animal and human), antler or tusk, peat, chemically unaltered clam or mussel shells, and animal and human bones all contain enough C^{14} to allow them to be dated.

The ideal material for radiocarbon dating is wood charcoal, burned at the time the site was occupied. As we dig, we look carefully for remains of ancient people's fires to give us samples for dating. At Koster we have been extremely fortunate, since the occupants of the various horizons left behind very clear evidence of their fires, complete with carbonized bits of wood, thus making it easy to establish dates for almost every horizon.

The question of when the first humans lived in North America has fascinated archaeologists for a long time. They have been excavating in the New World since the early 1800s, and during that time they have dug up thousands of sites. They have put together a broad picture of human occupation of North and South America stretching back to as early as 15,000 B.C., perhaps as early as 30,000 B.C. But the record is an uneven one, with large gaps in some periods. As new techniques are tested and used, knowledge of prehistoric North, Central, and South America is constantly being refined.

Until 1927 archaeologists thought that people had been in the New World for only a few thousand years. In the spring of 1925 George McJunkin, a black cowhand, was following the track of a lost cow near the small town of Folsom, New Mexico. As he looked up at one point, something caught his eye on the other side of a dried-up stream bed. He realized there were some bones sticking out of the bank and went over to examine them. He was puzzled because they seemed to be much larger than cattle bones. He pried parts loose with his pocket knife and took them along.

Word of McJunkin's find somehow reached J. D. Figgins, director of the Colorado Museum of Natural History in Denver, who asked to see the bones. He was able to identify the bones as belonging to a bison of a species that had become extinct about ten thousand years ago.

Figgins was excited at this find, and the next summer, 1926, he went to the site to excavate there. By 1927 he had found bones and projectile points together, in such a manner that there was no mistaking that the bison had been killed by ancient hunters. He named the points for Folsom, the nearby town.

This was electrifying news, for it indicated that people had been in North America much earlier than had been presumed.

Archaeologists in Figgins' time did not have any way of dating what they found, with any degree of accuracy. They could only date items relatively, by dating the geological strata in which they had been found.

A few years later, in the 1930s, projectile points were found near Clovis, New Mexico, which were fluted like the Folsom points but were considerably larger, as long as five inches. Clovis points eventually were dated to 9500 B.C., and Folsom points to about 9000–8000 B.C.

In recent years archaeologists have unearthed evidence that suggests that people were moving into widely scattered areas of both North and South America long before 9500 B.C. In 1970 Richard S. MacNeish, director of the Robert S. Peabody Foundation for Archaeology in Andover, Massachusetts, found a succession of stone-tool types in Flea Cave in the Ayacucho Valley in Peru. He was able to establish a C^{14} date for the earliest of these stone tools at about 20,000 B.C.

Since the assumption has been that early people arrived in the New World via the Bering land bridge (during periods when waters in the Bering Straits between what are now Siberia and Alaska receded) or across the waters of the Bering Straits and eventually reached South America, where is the firm *datable* evidence that they had passed through North America before reaching Ayacucho in 20,-000 B.C.? MacNeish's discoveries have stirred much interest among archaeologists working in the New World who believe that eventually they will find proof for earlier migrations of people across North America.

The earliest datable proof of people in the vicinity of Koster was the discovery of Clovis projectile points, like those from the Southwest which have been dated to about 9500 B.C. This indicates that Paleo-Indian people were in the Middle West at least eleven thousand years ago. Many Clovis points have been found in Lowilva (Alec has found some) but none as yet at the Koster site.

During the 1970 season we carefully collected carbon samples, although we had no funds to pay for having these analyzed at a special laboratory. One day toward the end of summer I received a call from the Illinois State Geological Survey in Urbana, offering to analyze them for free. We carefully packed our samples and dropped them off.

Just before we sent the samples to ISGS, Ira Fogel, a geologist who was doing soil research at Koster, found cultural debris in another layer below Horizon 10. So now we knew that, buried beneath the cornfield, there were at least eleven prehistoric occupations. The question tantalized us as we went about our chores in those last few weeks of the 1970 digging season, how old was this site?

5

The Kingdom of Lowilva

Alec led me to the Koster cornfield in the late summer of 1968, and we conducted our first full season of excavation there in the summer of 1970, but it was not until the spring of 1971 that we knew exactly how old the Koster site was.

That spring, a letter arrived from the Illinois State Geological Survey. I tore open the envelope.

The ISGS laboratory in Springfield had subjected the bits of charcoal from the Koster horizons to radiocarbon testing to determine when people actually had lived there.

Just reading the radiocarbon dates was enough to lift my spirits. It confirmed my original hunch that people had lived at Koster over a very long time span. Now we knew that prehistoric people had lived there, off and on, for almost eight thousand years, from before 6400 B.C. to A.D. 1200. It also confirmed my guess that they had lived there during the Archaic period.

DATES OF THE
KOSTER HORIZONS

HORIZON 14   ?
HORIZON 13   7500–6700 B.C. (Early Archaic period)

HORIZON 12

HORIZON 11   6400 B.C. (Early Archaic period)

HORIZON 10   6000–5800 B.C. (Early Archaic period,
             beginning of Middle Archaic period)

HORIZON 9

HORIZON 8   5000 B.C. (Middle Archaic period)

HORIZON 7

HORIZON 6   3900–2800 B.C. (Middle Archaic period)

HORIZON 5

HORIZON 4   2000 B.C. (Late Archaic period)

HORIZON 3   1500–1200 B.C. (Late Archaic period)

HORIZON 2   200–100 B.C. (Early Woodland period)

HORIZON 1   A.D. 400–1000 (End of Middle Woodland
             period, and Late Woodland period)

HORIZON 1   A.D. 1000–1200 Mississippian period)

The blank spaces in the list above indicate that, as yet, we have been unable to fix a firm date for the time those horizons were occupied because we have found neither enough carbonized material in them for dating nor diagnostic artifacts which could be dated from other sites. At the time of the first ISGS radiocarbon dating of Koster horizons, we had discovered eleven horizons. Since then, we have found Horizons 12, 13, and possibly 14.

More than six thousand years before Christ was born in Bethlehem, at least four thousand years before Stonehenge was constructed in southwestern England, and more than three thousand years before the great pyramids were erected to honor the Pharaohs in Egypt, people had settled in the great river valleys of the American Middle West, and now we were uncovering their settlement by chance.

I had found Lowilva by accident, because I am an incurable daydreamer at the wheel of a car. En route from

Evanston to St. Louis to give a lecture before a Rotary Club, I took a wrong turn.

At the time, in 1958, I was a graduate student at Northwestern University, working on a master's degree in anthropology. I had chosen Northwestern so that I could enroll in the African studies program, under the renowned anthropologist (the late) Melville J. Herskovits, and Creighton Gabel, now at Boston University, an archaeologist who specializes in east and south Africa.

I had received my B.A. in English literature from Dartmouth College, with a minor in anthropology, in 1953. After college, I had drifted from job to job, working first as a cab-driver, then in the steel mills, undecided about a career. I had a deep desire to practice archaeology, but could not resolve my inner conflicts about its usefulness to society. I am descended from Scotch-Irish settlers from Kentucky, Connecticut Yankees, and German-American industrialists. Although my parents encouraged my two brothers and me to choose our own careers, the deep, underlying utilitarian philosophy with which they had imbued us could not be shrugged off easily. The message had been clear: Choose what you want to do, but you must be a productive citizen, so do something useful. For more than five years I had struggled with the question, Is archaeology of any practical use to anyone?

In 1955 I had drifted to Cleveland and worked for a packaging firm. In Cleveland I met Ross Widen, an art dealer with a deep interest in primitive art. We formed a partnership; I would go to Africa and search for sculpture, and Ross would sell what I found.

For several months I traveled up and down the west coast of Africa by car, boat, and foot, penetrating small villages in the interior. My travels took me to Liberia, the Ivory Coast, Guinea, and Senegal. As I traveled, I became fascinated by the prehistoric ruins I encountered, none of which had been explored. Useful or not, I felt I must become an archaeologist and return to Africa to excavate prehistoric sites.

In 1956, after my return from Africa, I enrolled at Northwestern University to study archaeology. To earn money, I became a lecturer for the Redpath Lecture Bureau in Chicago, and it was on an assignment for this agency that I became lost en route to St. Louis.

I never look at road maps until I get lost. When I consulted a map on this occasion, I figured the fastest way to my destination was to drive south on Illinois Highway 100 to the tip of Calhoun County, where I could take the Golden Eagle ferry across the Mississippi River to St. Louis.

As I approached the town of Kampsville, driving south along the base of the western bluffs of the Illinois River Valley, I saw on my left several very large earth mounds. I was fascinated; from their size and shape, I knew they were not natural; I suspected they were Amerindian burial mounds.

I stopped and went over to examine them. There were seven loaf-shaped mounds, six on the east side of the road and one on the west. I climbed to the top of the one nearest the highway on the east side to look around. From there, I could see that they had been set in a roughly rectangular configuration.

The one I stood on appeared to be about 250 feet long and about 120 feet wide. Someone had sheared the top of it off with a bulldozer; there was also a jagged hole in the side. (Later I learned that the mound had been more than 20 feet high before the top had been bulldozed off.) It was obvious that someone was tearing the mound open to see what was inside.

The largest of the mounds, which had not been touched, appeared to be about 325 feet long by 135 feet wide and 25 feet high.

I drove into Kampsville, two miles south, and asked at Capp's General Store if anyone could tell me who was digging the mound. The clerk suggested I call on Kermit Suhling.

Kermit is a farmer and has lived in Kampsville all his life. He has been collecting Amerindian artifacts for more than fifty years, since his teens. He showed me his collection of artifacts and told me about the mounds. As we sat and chatted at Kermit's kitchen table, I felt very much at home. Kermit and I have been friends since.

"There's a whole series of mounds along there called the Kamp Mounds, after old M. A. Kamp, who founded Kampsville," he told me. "While Kamp and his son, Joe, lived, they wouldn't let anyone touch them. But after Joe Kamp died, the Kamp family, wanting to open up more farmland, decided to bulldoze the trees off most of the mounds, and in the process, Pete Kamp, M.A.'s grandson, decided to dig into one of the mounds to see what, if anything, was inside."

There had originally been ten mounds, explained Kermit, but when the state of Illinois put in Highway 100, it had been necessary to bulldoze three of them to make way for the road.

Later I learned that when the state had built the highway in 1929–30, the engineers realized they were going to have to destroy the three mounds, so they had notified archaeologists at the University of Illinois. The archaeologists conducted a brief salvage excavation and left a record of it. They had found stone crypts, which upon excavation were dismantled to make way for the highway.

I called on Aloysius (Pete) Kamp, at Kermit's suggestion, and explained my interest in the mounds as an archaeologist. He was very gracious and agreed to let me bring some students down to conduct an exploratory dig of the mound he had already cut into.

Back at Northwestern after my trip to St. Louis, I persuaded several college and high school students to return with me to the valley, and we spent a week excavating in the mound.

I returned again in the summer of 1959 with a small group of students, and although we had almost no funds, I managed to run a dig for several weeks.

Kamp Mound Number 9 (the mound we dug) turned out to be a two-thousand-year-old mortuary site of one of the prehistoric Hopewell cultures. Radiocarbon dating showed that it had been built (and people buried in it) between 50 B.C. and A.D. 150.

The name Hopewell is given to a group of contemporaneous cultures of the Woodland period, which flourished, as we have noted, during the period 100 B.C.–A.D. 450. They spread over a vast area of the eastern part of North America, extending from present-day Rochester, New York, in the east to Kansas City, Missouri, in the west and from the state of Wisconsin on the north to the Gulf Coast on the south.

Although they lived varying life styles over this large area, the Hopewell people shared a set of common trade goods and common styles in ceremonial artifacts. They made extraordinarily beautiful objects, which we today consider works of art. Many of these items, designed for body adornment or personal use, were symbols of rank and status and were exchanged among the elite. Among the beautiful objects the Hopewellians produced were pipes, breastplates, figurines, and earspools (circular objects worn in the ear as jewelry, as we wear earrings).

Kamp Mound Number 9 contained a Hopewell tomb in which had been buried five persons, including four adults and one infant. The latter had been placed on the knees of one of the females.

The tomb was rectangular, set in the middle of a circular earth mound. The tomb itself had been built of logs, which had long since crumbled and decayed, but we could still see the impressions of bark in the ground. The entire log structure had been covered over with a layer of limestone slabs. On top of the limestone slabs the Hopewellians had poured earth in variegated colors and textures to form a dome-shaped mound. Placing different-colored earths in the doughnut-shaped embankment surrounding the tomb may have been part of a burial ritual.

The chest of one woman was covered with quantities of freshwater pearls and cut-shell beads. The way these were arranged suggested that they had been attached to a garment, now deteriorated. As I looked at the carefully placed pearls and beads, I experienced one of those moments that occasionally come over every archaeologist of feeling intensely the common humanity we share with the people whose past we have discovered. The woman apparently was among the elite in her society. Her garment must have been made very painstakingly to achieve that effect, and I suspect it was a sign of her rank or wealth or both. Pearls are very rare in freshwater clams, and it would have taken an incredible amount of human energy to gather that many.

We also found hundreds of flint knives in the nearby village site, but almost no other artifacts. When I looked at this collection of knives, I wondered whether the Hopewellians had been cutting themselves at the burial site and had left the knives there as part of a ritual. Anthropologists have observed the custom of self-inflicted pain (by piercing, or cutting the skin) among various nonindustrial, tribal cultures as part of their mourning rites.

I used my report on Kamp Mound Number 9 as the basis for my master's thesis at Northwestern University, and then moved to the University of Chicago. By this time I had become intensely interested in the new theories and methods being argued about by archaeologists, and I wanted to join in on the discussion that was taking place at Chicago.

In the department of anthropology at the University of Chicago from the mid-1950s through the mid-1960s there was one of those rare combinations of teachers and students whose interactions would eventually have a major impact on archaeology. Among the teachers were four archaeologists whose ideas provided great stimulation for their students.

Robert Braidwood was tracing the origins of agriculture in the Near East. He was among the first archaeolgists to invite biologists and earth scientists to join his excavation team in Iraq and Iran and to conduct comprehensive studies of ancient people and their environment.

Robert McCormick Adams was using the concepts of anthropology to study the high civilizations of the Near East and Mexico, seeking an explanation for why these civilizations rose when and where they did. He was exploring the processes that give rise to urban versus nonurban states.

F. Clark Howell was involved in East African studies in the early evolution of near-humans and the beginnings of culture in the early parts of the glacial epoch.

Lewis Binford was interested in developing general models for explaining culture change and archaeological research strategies by which to test these models. He was trying to establish a genuine scientific method as the basic working principle for archaeologists.

Today, the group of graduate students who were influenced by one or more of these scholars reads like a roster of those in the forefront of experimentation in the new archaeology. Among them are Melvin Aikens, James Brown, Jane Buikstra, Kent Flannery, Leslie Freeman, John Fritz, Margaret Fritz, Richard Gould, James Hill, Frank Hole, Richard Klein, William Longacre, Thomas Lynch, Christopher Peebles, Fred Plog, Charles Redman, Patty Jo Watson, Robert Whallon, Jr., and Henry Wright.

The whole time I was working at the University of Chicago for my doctorate, and when I joined the Northwestern University faculty, I continued to excavate in Lowilva in the summers. Each year I managed somehow to scrape up enough funds barely to see myself and a small group of students through an excavating season. For several summers our little group lived in a series of dilapidated houses in and around Kampsville whose main appeal was that they could be had for very little money—in some cases, rent-free—because of their poor condition.

I must confess that, in our earliest days, along with teaching our students archaeology I instructed them in the art of living in the country on no money. One trick was to "hitchhike" corn.

Most of the corn grown in Lowilva is field corn, a hybrid grown for feed for cattle and hogs. In August it is in the roasting-ear stage and is as delicious as sweet corn if it is picked at just the right time. That summer we developed a strategy for obtaining fresh corn free.

We would drive around and find a cornfield in which the corn silk was just starting to turn brown, which indicates that the kernels are largely sugar. (Within ten days the corn turns yellow and starchy and is hardly edible.) We would park the truck near the field, and I would go into the field, where I was not visible from the road since the corn was so high. I would toss ears of corn out toward the truck, where the others would be waiting to retrieve them and put them in the truck.

Gradually we began to attract an informal support system among our families and some of the local farmers, after they had quietly observed our moneyless state.

My mother brought us boxes of canned goods.

On Sundays Mr. and Mrs. Willis Perkins, farmers from Toulon, Illinois, would come to visit their son, Ray, one of my students. after their first visit, they would arrive almost every Sunday with a cooked dinner—a standing rib roast, fried chicken, or baked ham. Thoughtfully, they always brought enough for us to have leftovers too.

Kermit Suhling, the Kampsville farmer who had become my friend, would call and report that his green beans were ready to be picked. For every two bushels we picked, we could keep one.

Sometimes we had to be inventive about our living arrangements. Once we lived in a farmhouse south of Kampsville which we rented for twenty-five dollars a month. There were seventeen of us. The seven-room house, which sat out among cornfields, was multifunctional—it served as dormitory, kitchen, dining hall, laboratory, and classroom. A hand pump in the back yard provided water for drinking and cooking, and there was the usual privy out back.

But there were no facilities for bathing. At the end of a day's work digging holes in the ground, one likes a wash, so we rigged up two outdoor showers; one for men and one for women. The showers were open-air enclosures made of tarpaper wrapped around railroad ties placed upright in the ground. They resembled French *pissoirs*, except that on top of each one was a barrel sitting on two-by-four planks and equipped with a spigot.

Each morning before leaving for the field we would pump water from the well and fill the barrels. While the sun was warming us at the site, it was also warming the water in the barrels. Unfortunately, if it was overcast the water remained cold and we ended the day with a cold shower.

As I stood in one of these makeshift shower stalls at the end of a hot day, I would look out over the top and think how lucky I was to be able to watch corn grow while taking a shower.

In 1960 A. R. Kelly of the University of Georgia and Joseph Caldwell of the Illinois State Museum planned a seminar on the Hopewell cultures. Archaeologists talk about the Hopewell people as easily as we talk about our neighbors, since that group of cultures is as well known as any archaeological phenomenon in North America. Kelly and Caldwell felt it was time for a stock-taking seminar, to pool knowledge gained by several archaeologists and, from that, possibly to redefine our views of Hopewell cultures. They invited me because of my work on Kamp Mound Number 9.

Spurred by this seminar, I began to research more broadly what was known about the Hopewell in the Illinois and Mississippi River valleys, and from then until 1970, when we began a large-scale research effort at Koster, I concentrated on that culture.

Through the 1960s I excavated several Hopewell village sites, including the Apple Creek and Peisker sites. The Hopewell in Illinois chose to locate their villages at the base of the bluffs, usually at the juncture of the tributary valleys with the main river valley. In Lowilva they also established some villages out on the floodplains, on what had been beaches, because either the Illinois River or tributary streams once ran there.

As I continued my work on the Hopewell cultures, I observed that there appeared to have been a dramatic increase in the complexity of social systems in the American Midwest in about 100 B.C. The Hopewellians had developed elaborate mortuary practices, a complex social system with power concentrated in the hands of the people at the top, and an extensive trade or exchange network. Concurrently, they had gone through a flowering of creativity, producing artifacts of distinctive artistic styles.

As I examined the evidence, I wondered why these dramatic changes had occurred only in the major river valleys of the Midwest and not in contiguous areas, such as the upper prairies, rolling forests, or minor river valleys? And why did there appear to be a northern barrier to the occurrence of the phenomena? Similar changes did not appear to have taken place north of central Wisconsin or central Michigan. Had the climate influenced these cultural changes? Might they be explained in terms of some subtle series of relationships between people and their environment? But how to go about studying these?

When I was a youngster, my father, Carl Struever, used to take me across the street from our home to visit the plant of the American Nickeloid Company, which had been founded by my grandfather, Rudolph Struever, and W. H. Maze in 1896. My father was then general manager of the company. I would ask him what each of the men working there did, and carefully he would explain their jobs to me. It was deeply instilled in me that any human endeavor has to be a cooperative venture of specialists working together.

These impressions from my boyhood, combined with what I was hearing from the scholars in graduate school and my own speculations about the Hopewell phenomena, led me to realize that we were going to have to shift to team research, to organize groups of archaeologists and environmental scientists to work together over long periods of time if we were to find the answers to all these important questions.

And gradually it dawned on me that I had accidentally chosen an area that was admirably suited to experimentation in new techniques in archaeology. I no longer felt I had to go to Africa; I had found my research universe in my own back yard.

The Lower Illinois River Valley is ideal for practicing the new archaeology.

First of all, it is rich in archaeological sites. Three major rivers—the Mississippi, the Missouri, and the Illinois—meet at the base of the valley. Archaeologists have long known that major populations of prehistoric Amerindians lived in the great river valleys of the Midwest, because these were areas with great concentrations of plant and animal life, and people who followed a hunter-gatherer way of life found it easy to make a living.

Lowilva has had a long and elaborate cultural prehistory, which has not yet been disturbed by modern industrial life. There are many thousands of prehistoric sites in the United States, but not many places where there are hundreds of sites, still intact, which permit archaeologists to get a total picture of how human cultures adapted to their landscape.

As new archaeologists we study what we call subsistence settlement systems. We are interested in examining several different types of sites which may have been occupied by the same people, such as a village, hunting camp, or food-processing camp. Consequently, we have staked out, as our research universe, a 3,200-square-mile area of Lowilva, running eighty miles north to south on the Illinois and Mississippi rivers by forty miles east to west. It is bounded on the south by Alton, Illinois (just north of and across the Mississippi River from St. Louis), on the north by Beardstown, Illinois, on the west by the Mississippi River Valley, and on the east by Interstate Highway 55. We set those particular boundaries to include a geographic frame which would have been large enough to include all of the potential resources which might have been exploited by any of the cultures we want to study. We consider that the time span we are studying ranges from about 10,000 B.C. to A.D. 1673.

Kampsville appealed to us as a headquarters because it is well-situated within this 3,200-square-mile area. It also is ideal for our purposes because of its isolation from the university campus and other competing activities. Scholars from many disciplines and students mingle together daily; for several months each year we become an archaeological community. On a large campus this would not occur.

Kampsville is a ninety-minute drive from downtown St. Louis, which gives us easy access to tools, chemicals, and other supplies available in a major city, and a five-and-a-half-hour ride from the Northwestern University campus in Evanston, a suburb of Chicago. At early 1960s prices, it cost the expedition about twenty dollars to transport a student or staff member from Evanston to Lowilva and back, as compared to several hundreds of dollars which it would have cost us to transport the same person to and from a site in a foreign country. The location also makes it easier to recruit the scientific experts needed to undertake major research. Many scientists might find it difficult to arrange to spend three months abroad, for several years in a row, whereas they can come and go to and from Lowilva fairly rapidly and inexpensively.

For all of these reasons, Kampsville seemed an ideal setting in which to develop a permanent archaeological research and teaching center.

In addition, because the area has a long tradition of scholarly studies of the environment, there is a rich storehouse of information on the environment in Lowilva such as would be totally unavailable in many parts of the world. At the University of Illinois in Urbana there are a number of masters' theses in the zoology department on animals that are native to the region. Preliminary studies have been done by scientists working on the river valley ecology. And the records of the U. S. Government Land Surveys, made with great care in the period 1815–20, are available in the Illinois State Archives in Springfield. These give very detailed descriptions of the vegetable cover in Lowilva, and show where different plant communities were located before the first Euro-Americans began to move in and change the landscape for agriculture. These botanical and zoological studies give NAP scientists a wealth of information to draw upon in their attempts to reconstruct the environment of ancient people throughout the area.

I think it is very fitting that I should be excavating in Lowilva for personal reasons, too. Like the prehistoric people who lived in Lowilva, I, too, am a Midwesterner. And like them, I grew up on the banks of the Illinois River, in Peru, Illinois. Our family home in Peru is on top of the bluff, and my room had one window. Every morning for eighteen years, after I woke up, I jumped down from the upper bunk, and as my feet hit the floor, I would look out the window at the Illinois River. I still love to observe the river, from the second-floor windows of my office in Kampsville or from the bluffs overlooking the valley. I spent my boyhood hunting for small game, and fishing in the floodplain lakes of the valley just as prehistoric Amerindian boys did, no doubt, centuries before. They had one advantage over me—when they lived there, the river was clean enough to fish in. In my boyhood it already had become too polluted for that.

And it was in Illinois, not far from my home in Peru, that I had my first exposure to archaeology, when I was nine.

When we were growing up, my brothers, Rudy and Carl, and I visited the American Nickeloid Company factory frequently. One day in 1940 I walked by the desk occupied by Henry Decker, an administrator of the company. There on Mr. Decker's desk, I saw a shiny black stone axe; it was made from basalt, fully grooved (meaning the groove went all around the stone) and beautifully carved.

I was fascinated; I stood and stared at it. Finally I asked Mr. Decker, "Where did you get that?"

Mr. Decker explained that he found the axe in a field out near the old coal mines near Cherry, Illinois, and that it no doubt was from prehistoric times.

"Would you take me there sometime?" I asked.

Not long after, on a mild, sun-washed April day, Mr. Decker and I drove out to the mines. He had collected artifacts for a long time and knew there were prehistoric sites near there. We spent the day looking for artifacts. To my joy we found three tiny "bird" arrowheads (so called because they were so small it was thought that they were used only for birds). Today I know that those arrowheads were triangular type points from the Mississippian period, nearly a thousand years ago.

After a hard day of tramping about on the prairie, we sat down on a stump to rest. The stump was on the edge of a field of black dirt which had been newly washed by the spring rains. Suddenly we noticed a bunch of white flint chips scattered on the ground in front of the stump.

Pointing to the ground, Mr. Decker asked, "Stuart, what do you think happened here?"

"Well, wait a minute; let me guess," I replied.

"Look at these flint chips. What do they tell you?"

I stood in silence, staring down at the chips. "Well, I

don't know, but it looks like someone was making arrowheads."

Mr. Decker was delighted. He said, "That's right. It looks as if we have a place where an Indian worked. Maybe he was making the kinds of arrowheads we picked up today."

I bent over and picked up some of the flint chips and held them in my hand. I rubbed them and imagined how they had felt to that unknown American Indian long years before.

For a few moments I imagined that I could see the man as he sat at the edge of a hunting camp, swiftly striking stone to fashion an arrowhead, not noticing the flakes of flint that dropped to the ground as he worked.

It was one of the strongest emotional experiences in my life. From that time on, I wanted to become an archaeologist.

After that adventure with Mr. Decker I began to collect Amerindian artifacts. This wasn't difficult. The Illinois River is literally surrounded by archaeological sites; all one needs to do to look for artifacts is walk through a cornfield.

My Struever grandparents lived a block from us, and they had a large enclosed porch on the second floor of their house. They turned this over to me, and there I set up a small museum, displaying my artifacts on card tables.

When I was fifteen, I used to borrow the family car and with a friend, Ken Charlton, would drive to a farm my family owned south of the adjoining town of LaSalle, in the Vermillion River Valley. Ken and I would walk about the farm looking for pieces of worked flint and arrowheads. We became so engrossed in our hobby, which we carried on for at least two years, that we began to catalogue the pieces we collected and log them on a map. Eventually we made an informal survey of a ten-mile stretch of the Vermillion River Valley. It was not until I began to study archaeology that I learned there is a name for what we had been doing—surface surveying.

During the summer of my freshman year at college, when I heard that archaeologists working for the State of Illinois Department of Architecture and Engineering were excavating an old French fort, Fort St. Louis, on top of Starved Rock, between LaSalle and Ottawa, Illinois, I badgered them into hiring me as a laborer.

Later, on one of my trips as a lecturer while I was a graduate student at Northwestern, I met an extraordinary man, Father Francis Borgia Steck of Quincy College, Quincy, Illinois. Although Father Steck lived a cloistered life as a Franciscan monk, he was well known as a Franco-American historian. At the time we met, he was a very old man, encumbered by many physical problems, but he was still very active mentally. Some of the younger priests and his friends among the laity helped him, and I was delighted when he asked me to assist him too. Several times I was able to find material for him in libraries in Chicago. And over a period of months I spent several weeks as his guest, working with him on a daily basis, rewriting material for him, criticizing a manuscript for clarity, and generally trying to be helpful.

Father Steck was reanalyzing the Marquette and Jolliet exploration of North America in 1673–74, using copies of primary documents. Through his own research and that of colleagues he had found some primary material that had not previously been studied by historians. As he worked, he had found increasing evidence that suggested that Marquette probably had not been on that particular trip: there seemed to be no primary documents supporting his actual participation in the voyage; accounts of the voyage appeared to have been based on hearsay.

Father Steck's position was not a popular one. Marquette, who had been a Jesuit, had become a hero for exploring unknown territory in the New World. Yet Father Steck persevered, convinced that it was important to document in a careful, scholarly way what appeared to be a contradiction of previously accepted facts.

When I entered Father Steck's world, I was able to observe a mature historian writing his most significant work. Yet he was dealing with a very narrow topic. I was struck by this, and by his dedication to searching out facts and his reverence for the truth. I had never been close to someone who held that kind of love and respect for the search for truth, and it affected me profoundly.

I became aware that it was out of such dedicated, intensive, narrow searches for facts that greater truths could be built. And I recognized that an archaeologist doesn't have to understand all of the prehistoric cultures of the world, but can make a contribution by studying one culture thoroughly. Whatever ambivalence I had about a career as an archaeologist disappeared during my friendship with Father Steck.

Fatefully, about that time, I started out for the Rotary Club in St. Louis, daydreamed at the wheel of the car, became lost, and took Illinois Highway 100 into Kampsville.

I could not know it then, of course, but as I scrambled to the top of Kamp Mound Number 9, my entire life was to be overwhelmingly influenced by that moment and that act. I was never to return to Africa to excavate the ruins that had so intrigued me. I was to choose, instead, to spend my professional life as an archaeologist in an area that eventually lured me much more strongly than Africa had—the Lower Illinois River Valley, or, as I like to call it with affection, the Kingdom of Lowilva.

6

"The Big Hole"

Stuart, that's a big hole!"

Lewis Binford, professor of anthropology at the University of New Mexico, and an old friend, had come to Kampsville to be guest lecturer at the NAP field school. At the moment, he and I were standing at the edge of the Koster site.

It *is* a big hole.

From the handful of six-foot squares we first opened in 1969, the Koster site gradually has been expanded to become a huge, L-shaped excavation. The site measures a hundred feet long on the east wall, eighty-five feet long on the west wall, seventy-two feet long on the north wall, and forty feet long on the south wall. At its deepest, it is thirty-four feet below ground level. The hole varies in depth at different points, partly because of the different rates at which we dig sections and partly because for safety we have had to step-trench it as we have dug down.

As I stood there with Lew, I found myself thinking ruefully of all the problems, technical and financial, we have been forced to deal with because of that enormous hole in the ground.

Just the past week, as I sat over an early morning cup of coffee in my office, watching the Kampsville ferry plying back and forth, the phone had rung at 6:30 A.M. It was Gail, who had just arrived at the Koster site with the crew.

"Stuart, the north wall has collapsed. It's threatening to completely cover the area where we are digging Horizon 11. What shall we do?"

A few moments later I drove my car onto the ferry and for once was impatient with its methodical, plodding pace across the narrow span.

At the site, I joined Gail down at the thirty-two-foot level and inspected the damage. It had rained heavily during the night, and with no one there to remove the rainwater as it filled up the hole, the inevitable had happened. A portion of the north wall had slumped at the bottom, depositing wet mud over part of Horizon 11. The rest of the wall was holding, but for how long?

We decided that temporarily we would shore up the wall to keep the rest of it from collapsing. We had the crew drive sheets of metal into the ground to surround the slumping wall.

Ever since we have been digging at Koster, we have struggled with problems of water from two sources—rainwater and below-ground water. Each, in its own way, has been difficult, and expensive, to deal with.

Of course we had known from the beginning that digging a deeply stratified site would present special problems.

When you dig a site like Koster, where you are trying to recover accurately the remains of as many as thirteen different villages, you must of necessity take a long time. This means that earth walls have to stand open to the elements for several years. Unfortunately, rainwater collects very readily in that great open bowl in the ground, and if it is allowed to stand, it can cause the walls to weaken and collapse. The loess which covers most of Koster is treacherous in its nature; it can stand for very long periods, and then suddenly collapse. Other soils will peel off horizontally, but loess has what is called vertical cavage—it drops suddenly, without warning.

Fortunately, we had one of the world's foremost experts on soils right on the Northwestern University campus. Jorj Osterberg, chairman of the Department of Civil Engineering at the Northwestern Technological Institute, had visited Koster and advised us to step-trench our excavation. This means that instead of digging straight down and ending up with thirty-four-foot-high standing walls, we staggered the levels as we went down, so that we had shorter walls. To do this, we had to start the sides of the excavation quite a ways back from what we thought would be the center. Jorj had suggested that we have standing walls no higher than ten feet, and that each horizontal level be at least ten feet wide. This was not always possible to achieve, since there were portions of the site where we knew there was valuable data to be retrieved, and we had to dig there, so occasionally we have had taller standing walls.

I called Jorj and asked his advice about the collapsed wall. He suggested that we extend the outer margins of the entire site, starting at the top, and step-trench again for safety. So once more our bulldozer and end-loader were put to work, and the Koster hole became even larger.

Aside from Jorj's recommendations on step-trenching for safety, there is another reason for continually extending the outer boundaries of a site like Koster. We are digging at least thirteen different sites, one on top of another. The villages were not settled right on top of each other; their boundaries vary all over that hillside. And as new archaeologists, we are looking for large amounts of data for at least a dozen scientists. Consequently if we want to collect meaningful samples of material from the most deeply buried village, we need plenty of room down at the bottom in which to excavate. We had observed other deep sites, where they had not had funds for machinery to remove enormous amounts of dirt and had ended up digging in a postage-stamp-sized space in the bottom horizon. We didn't want that to happen to Koster.

As time went on, it became apparent that we must take other special steps to protect the exposed earth walls from rainwater. We covered the entire surface of the site with enormous sheets of black plastic. We try to keep these in place by hanging old automobile tires as weights down the sides of the walls over the plastic. This is a very expensive operation. When the plastic blows off in a violent wind, which it does repeatedly, it must be replaced. We have several people who spend their time perpetually redraping the plastic sheeting and moving the old tires to keep it from blowing away.

We also have had continuing problems with ground water, which has impeded digging and threatened to weaken standing walls. We first encountered problems with ground water at Horizon 8 but were able to relieve them by installing six de-watering wells at the edge of the site.

However, when we began to work in Horizon 12, about thirty-two feet below ground level, the de-watering wells did not remove water fast enough if the diggers were working in an area some distance from the wells. They found themselves having to scoop up thick mud. So once again we turned to our de-watering experts and, under their supervision, installed small wells right next to each area in Horizon 12 where excavators would be working.

Back in 1971, in response to my request for help, Alec had found an old corn elevator for two hundred dollars and adapted it for conveying back dirt out of the hole. We called it the Helton Archaeological Earth Mover (HAEM), and on hot, sticky days weary students bent over their shovels were grateful to Alec when the noisy, clanking conveyor chugged along, carrying out dirt.

Unfortunately, the HAEM was so noisy that one had to shout to be heard over it, and it is essential for diggers to be able to confer continually with the supervisor at a site. So we had long since replaced Alec's original corn elevator with a quieter, more efficient machine for removing dirt. Alec had converted this model from a fertilizer conveyor. Although we expected the Helton Archaeological Earth Mover II to last for only a year, it's still in operation, six years later. In all that time it has saved us thousands of hours of human drudgery that would have been required to remove that much dirt from the excavation. And because of Alec's care in designing it, no one has been injured by it.

As Lew and I descended into the big hole and walked about inspecting the excavators' progress, we kept up a rambling conversation, as old friends will. Lew is considered the leading theoretician of the new archaeology; he loves to talk about how to bring about changes in the way archaeologists work.

Despite the theoretical changes taking place in archaeology, many archaeologists have failed to cope with some of the discipline's most basic practical problems. American archaeologists, living in the midst of the most highly developed technological culture in the history of the world, continue to operate as diggers as if they were still in the early stages of the industrial revolution.

If you were to visit many of the sites being dug today, you would be astonished at the amount of soil being removed by human labor. Most excavators still use student power and wheelbarrows to remove back dirt. Possibly twenty-five per cent of a student's time at such sites is spent removing dirt.

With the new archaeology, the labor can only increase. Sampling requirements of the new archaeology are stringent and call for the removal of massive quantities of material. To expedite their work, archaeologists could use many different types of earth-moving equipment such as bulldozers, end-loaders, back hoes, road graders, and trenching equipment. No single archaeologist can afford to buy such equipment, nor could most universities; it costs too much to buy, operate, maintain, and store these machines.

But it isn't only money that keeps some archaeologists operating out of the mainstream of twentieth-century America. Part of the reason they are so lagging in applying modern industrial machinery to their own work is that many of them view their discipline as an intellectual pursuit, and they remain oblivious to the many time-, labor-, and money-saving devices they could use to advantage. Many seem content to keep their technology to the whisk-broom and dental-pick level and to cling to a view of themselves as frontiersmen, bravely going it alone with a little band of followers.

Even we at NAP, who pride ourselves on our ownership of two earth-moving machines, could improve our efficiency if we had more. A few weeks before Lew's arrival, I had an experience that once again reminded me how far behind we are technologically.

I had given a talk before a group of engineering and mining specialists at the University of Missouri in Rolla. Among the slides I had shown was one which pictured the conveyor system Alec had rigged up for us. I was proud of that conveyor; it showed that we were being pretty smart archaeologists and not carrying out all that dirt by hand.

In my talk I mentioned that I find it ironic that archaeologists come from all over the world to Kampsville to be introduced to the mysteries of the flotation method (a technique I developed for retrieval of small animal and plant remains). In terms of twentieth-century technological development, I consider the flotation process represents early industrial era behavior. The fact that I, who am totally untutored in engineering or mining technology, could have developed a method that is now world-famous merely reflects the relative lack of technological sophistication in archaeology.

After my lecture one of the engineers approached me and said in a sympathetic voice, "I see what you mean, Dr. Struever, when you say that archaeology is so far behind technologically." It took me a few moments to realize he was referring to our wonderful Helton Archaeological Earth Mover II!

But he's right, of course. Some day we hope to acquire more up-to-date equipment. In the meantime, we are grateful to those who have donated either machinery or the money to purchase it to NAP.

Back when Koster was first being explored, there were about fifty open squares across the cornfield. Beside each open hole sat a pile of back dirt. I looked at those piles and groaned at the thought of trying to remove all that dirt by wheelbarrow.

I went to several big-name companies which manufacture earth-moving equipment to ask for help, with no luck. Then I turned to a friend whose father, Dennis T. Buckley, was general manager of the J. I. Case industrial-equipment dealership in La Grange, Illinois. Dennis was from South Africa and had seen earth-moving equipment used at an archaeological site. Through his intercession J. I. Case agreed to give NAP a Case 350 end-loader, a small machine, ideally suited to rolling about the site to scoop up back dirt.

Case has continued to provide us with a machine at no cost for the past seven years; every two years the firm sends down a new model of the Case 350 and takes back the used one.

We also bought a second-hand bulldozer, which Jane and Greg have used to clear brush and to create temporary roads up to the mounds for their trucks, which carry equipment. (Jane drives the bulldozer herself, after watching a student back it up almost to the edge of the bluff crest.)

De-watering wells to control the water below ground level, conveyors for removing back dirt, and bulldozers to remove large amounts of dirt from a site all cost a great deal of money, more than any ordinary archaeological expedition is able to obtain. Each year since we have begun to dig Koster, I have been forced to spend less time as an archaeologist in the field and more time as lecturer out on the road to raise the funds necessary to keep us going.

On that day, as Lew and I moved over to examine an interesting feature in Horizon 11, Lew happened to glance up at the tourists who stood at the edge of the south wall looking down at us. There was a group clustered around Gordon Will, one of our chief guides, listening to his talk about the site. All of the men, women, and children in the group were wearing similar eye-catching hats. Each hat was crocheted of chartreuse yarn and had set into it four pieces of tin made from flattened beer cans (we learned this later from Gordon, who, on inquiring, had been told that they were a family camping group and that they had made the hats themselves). On each of the tin panels was emblazoned the name of the place the group had visited. One tin panel read "Disneyland," and another read "The Koster Site."

"Stuart, you're running a goddamned circus here," said Lew, laughing.

"No, I'm not," I replied. "I'm sharing knowledge with these people and they love it. They are fascinated to learn that North America has an impressive prehistoric past, just as you and I are. Besides, they are our lifeline."

And they are.

Back in 1970, when I was casting about for ways to raise funds to excavate Koster, I thought of all the people who have enrolled in our field schools or come to my lectures. I recalled their eagerness for any new information we could give them about their predecessors who had lived in America in prehistoric times. So I thought, Why not share the discoveries from this exciting site with more people? They could watch us work; we would explain what we were doing, and why. In return we would ask them to help support the expedition's work in recovering the American past.

So one day I asked Teed, "Is it O.K. if I invite a few people to come watch us work?"

"Sure, Stuart; ask anyone you want."

I sent word through the media that the public was welcome to visit the Koster site. We set up a small archaeological museum in Kampsville and persuaded Mrs. Marguerite Schumann, a science teacher at Calhoun High School in Hardin, to become curator.

Neither Teed nor I could have foreseen the results.

The first year, 1971, about ten thousand people came.

The second year twenty thousand people came.

And every year since, during the excavating season, more than thirty thousand people have made their way to Teed's former cornfield to learn first hand something about their country's prehistory.

To answer their questions, I persuaded three high school teachers—John Nelson of La Grange, Illinois, Gordon Will of Oak Park, and Peter Gilmour of Chicago—to come and serve as chief guides. To help handle the crowds, we recruited volunteer guides who work without pay (we give them room and board).

Our visitors come from nearby and from all over the world.

A busload of Japanese college students, none of whom spoke English, stood fascinated as they watched our semi-naked excavators at work in the big hole. Through an interpreter one asked, "How much are they paid to work so hard?"

When they learned that our students pay a large fee for the privilege of doing back-breaking labor under the broiling sun, murmurs of astonishment rippled through the group. (Actually, the students are paying for college courses they are taking in the summer field school. Their work at the site is an important part of these courses.)

The American people have responded magnificently to our appeal to help support the work at Koster and in Lowilva.

In 1977 our operating budget was over half a million dollars, and more than sixty per cent of that amount was provided by private citizens and private corporations. The gifts we received ranged in size from as little as a quarter to some amounting to several thousand dollars. We are the only archaeological program in the world whose work is supported by the public to such an extent.

We have, of course, received funds for operating from Northwestern University. We also have received large grants from the National Science Foundation for special projects such as computer research. And we have received grants from private foundations.

Back in Koster's earlier days, before we were widely known, frequently we found ourselves on the verge of closing down from lack of funds.

In the summer of 1972 we faced a financial crisis. The

bookkeeper informed me that we did not have enough cash to pay our bills after that week. We would need another eleven thousand dollars to see us through the season. I racked my brains, trying to think where I could raise the money. Then I put the problem aside and went off to greet a special guest.

Richard Ogilvie, then governor of Illinois, and his wife, Dorothy, had learned about Koster on television, and Mrs. Ogilvie came to see the site that August. As I showed her the excavation and took her on a tour of the laboratories, I mentioned that we were about to close down for the season because we had run out of money.

Mrs. Ogilvie was very upset when she heard this. "Dr. Struever, if you come up to Chicago for a day, I'll help you raise that eleven thousand," she said. "And if we can't raise it right away, I'll pledge it to you myself. You must not stop this important work."

Mrs. Ogilvie and I met at the Mid-America Club in Chicago for lunch. Irving Kupcinet had mentioned in his Chicago *Sun-Times* column that morning that we were to lunch there as we tried to figure out how to raise the money.

Mrs. Ogilvie and I had just sat down when a waiter came to our table, bearing a plug-in telephone.

It was someone offering to make a gift of a thousand dollars to help keep the excavation of Koster going.

Before we had finished lunch, enough pledges had come in for me to go back to Kampsville and tell the staff and the students that we could continue our work at least for the rest of that season.

Over the years, we have discovered that the old-fashioned custom of helping a neighbor in distress is still very much alive in the Middle West.

Our neighbors in Greene and Calhoun counties have shown great interest in our work at Koster. They call on us—farmers and townspeople, families spilling out of pickup trucks or station wagons. They chat, tell us about their crops or local business problems, and ask polite questions about what we are doing. Without any prying, they have been very sensitive to our financial state. When we have been hard up, our neighbors in Lowilva have responded by giving us what they could, frequently gifts of food and useful items.

One day as I emerged from the Koster hole and walked

down the slope to Teed's back yard, I encountered a farmer standing next to his truck. At his feet were several baskets filled with green beans, tomatoes, and squash.

"Dr. Struever?" he asked. I nodded, and we shook hands. Then, awkwardly, he gestured toward the baskets at his feet. "Thought you might be able to use these. Big crop; we've got more than we can use." Someone came up and spoke to me, and as I turned to acknowledge the newcomer, the big shy farmer hastily climbed into his truck and drove away before I could even get his name.

His generous act has been repeated many times, in varying ways, by other neighbors in Lowilva.

The parishioners of St. Anselm's Church in Kampsville donated use of their parish hall seven months of the year, for six years, for the expedition's dining and lecture hall.

Local farmers Donald Moss, Joe Brannan, Kenny Brangenburg, and John "Pete" Schumann (NAP's chief of maintenance) have fattened and butchered hogs and steers for us at cost.

One year the citizens of Kampsville held a fish fry and raised seven hundred dollars, which they donated to the "arkies."

And as I began to make rounds of corporations to ask for help, they, too, donated goods. Sears, Roebuck and Company earned cheers from the excavation crew when their gift of aluminum ladders was unloaded at the site (no more going up and down deep holes on ropes). Sears also donated house paint, which allowed us to transform dingy, worn buildings into cheerful hues of sunset gold and light elephant gray. Montgomery Ward and Company donated dressers for dormitories, brightening rooms which theretofore had contained only cots or old army beds. The Marquette Cement Manufacturing Company donated cement for building foundations. Container Corporation of America donated custom-made artifact boxes (we use an enormous number of these while excavating seven sites annually). Jewel Companies, Inc., and the Quaker Oats Company have donated considerable quantities of food.

In our third year of operation we formed a not-for-profit organization, the Foundation for Illinois Archaeology (now merged with the Northwestern Archaeological Program), and invited people to become members. In return for a contribution, they receive NAP's newsletter, *Early Man,* which gives them news of Koster and other sites we are digging as well as information on new developments in archaeology. Our foundation has a special category for corporate memberships; we have eighty-five corporate members, many of which had never before donated funds for American archaeology.

The generosity of the American people and of these private corporations has enabled us to make important changes in the way we operate. We have gone from a small group of archaeologists, struggling to practice an increasingly complex discipline with insufficient funds, to the formation of the Center of Archaeological Research,* whose efforts will be devoted exclusively to the support of experimentation in archaeological research and education.

For the past several years NAP has sustained the largest multidisciplinary team ever put together for archaeological research in North America. In 1977 the NAP roster included 265 people—including 12 scientists, 15 field supervisors, 218 college, high school, and junior high school students, and 20 administrative and maintenance workers. What this group has accomplished at Koster has changed our understanding of North American history.

* See Appendix.

7

Mary and Teed

The discovery of an important archaeological site in their cornfield, named for them, has brought fame to Mary and Teed Koster. They are proud that their name appears in several books and on a National Geographic Society map of pre-Columbian North America. But they are much too modest to talk about it.

Along with the fame, the excavation of the Koster site has brought the Kosters swarms of busy archaeologists, prying reporters and photographers, bossy television crews, and busloads of tourists. Every day during the digging season, about 350 people tramp through their back yard, park next to their hogpen, poke at their pigs, sit on their chairs, scramble over their farm machinery, and drink from their pump.

Mary and Teed have met the onslaught with incredible equanimity. Childless, for almost thirty years they shared a quiet life on their farm, watching the seasons come and go, participating in the social life of the farm community in Greene County.

Teed, a man who has had an intimate relationship with the soil for more than forty years, has no inclination to dig up the past. That he leaves to the "arkies," whom he tolerates with quiet amusement.

"No poor man's kid ever dug in that hole," he once told a newspaper reporter, thus revealing his own feelings about the fact that NAP students pay for the dubious honor of digging the hole in his cornfield ever deeper.

The world now comes to Teed's back yard. He has no desire to return the visit. During World War II he was in

service, and the traveling he did then, he feels, was enough
for a lifetime.

Now, all summer long the Koster back yard churns with
people coming and going. The Koster crew arrives at 6:30
A.M. The guides come at 9:00 A.M. And the public begins
to come at 9:30 A.M., when the site formally opens. They
all leave at about 5:00 P.M.

Sometimes people won't wait for the site to open. Teed,
coming out to do chores at 6:00 A.M., will find a lone visitor standing at the edge of the site, meditating silently in
the sun, off on some private trip back into the past. Teed
will greet the intruder politely, as he greets everyone,

"Morning," and then he'll whistle for Gypsy, his German shepherd, and proceed to his work.

In the midst of all the activities, the Kosters tend to
their daily chores as they always have, on the farm and in
the white frame house.

Teed, in his late sixties, is semiretired from farming; Joe
Brannan, Alec's son-in-law, and his son, Stanley Brannan,
work the farm for Teed. (The Brannans farm six contin-
guous farms, including Alec's.) But Teed still rises early,
and daily at 6:00 goes out to feed the hogs. Each morning, as the "arkies" spill from the tan NAP bus, they cross
paths with Teed on his rounds. Gypsy deserts her master
temporarily to greet the youngsters, then trots off to join
Teed.

On Mondays Mary hangs out the wash.

On Thursdays Teed drives Mary into town to the
hairdresser and to do her shopping.

But the invasion of archaeologists and curious tourists
has brought about some changes in their lives.

Late summer mornings, Teed, wearing a long-sleeved
white shirt and dark cotton pants, the signal that his work
day is over, comes out of the house and offers the St.
Louis *Globe-Democrat* to the volunteer guides sitting in
the glider under the big tree in back of the house.

The guides have worn the glider's cushions down to the
springs. When we noticed this, we arranged to have a
handsome new aluminum glider, with bright yellow-and-
white flowered cushions, delivered to the Kosters as a replacement. The new one sits in a place of honor on the
front porch of the farmhouse, and the guides continue to
wear down the springs on the old glider.

If Vern Carpenter (now seventy-eight and still working

in the big hole at Koster) has come up to rest for a while in the shade, Teed joins him for a chat.

Like many people in the area, Teed collects artifacts, and occasionally a privileged visitor is invited inside the farmhouse to view them. Teed pushes aside Mary's hand-crocheted doilies from the dining-room table and lays out stone projectile points, metates, and various kinds of knives.

Each day, promptly at noon, Teed disappears behind the back door of the house for lunch, which in farm country is referred to as dinner.

Frequently he emerges a moment or two later, bearing a white enamel basin filled with luscious-looking, ripe, red tomatoes, collected a few moments earlier from the Koster vegetable garden. Without a word, he deposits the basin on the old wagon-bed which serves as a luncheon table for the "arkies," smiles shyly, and retreats.

I had noticed that the size of the Koster tomato garden and the number of watermelons they grew increased shortly after the "arkies" began digging in the cornfield.

"We have more than we can eat," said Teed when he presented the first few basinfuls of tomatoes to Gail.

After washing the dinner dishes, Mary often joins the group of guides under the trees. But she doesn't sit still for very long; she is self-appointed traffic manager for the automobiles, trucks, recreational vehicles, and buses which pull in and out all day long. Suddenly she will dart up to direct someone. Her favorite targets are the large tour buses.

"No, no, not over the bridge," she calls to a bus driver. "You can pull up next to the shed," and she waves a hand in that direction. In her starched cotton housedress, apron, black oxfords, white bobby socks, and neatly waved gray hair, there is no mistaking who she is—the mistress of this small, busy corner of the universe—and she is seldom, if ever, challenged.

The Kosters seem to thrive on the enormous amount of energy that flows about them daily, as tourists come and go, and the "arkies" toil in the pit, or play, during breaks and lunch hours, in the Koster back yard.

Frisbees swirl over Mary's gray head as she sits chatting. Young "arkies" smeared with mud cluster around the pump and bend to souse their heads in it. Or they sit with their feet in the cement trough, splashing water over their legs, washing off eight-thousand-year-old dirt. From the hill on the east side of the house the soft, plaintive notes of a flute waft down as assistant supervisor Paul Farley puts in his daily practice stint.

A couple of students search the equipment shed and come out bearing kittens. In Kampsville dormitories the pet population threatens to engulf the student population. Many of the kittens are from litters born to the Koster cats. Mary is delighted when a student asks for a kitten; she keeps the new owners informed about the progress of each kitten until it is ready to be claimed.

When parents come to visit, students bring them over to meet a beaming Mary. Each year she remembers the old faces and learns the names that go with the new ones.

Frequently visitors seek out the Kosters. If the visitors are also farmers, hosts and strangers greet each other warmly and soon are launched happily into exchanges over the horrendous prices of fertilizer and the dismal prices for hogs.

A woman from the city, slightly hysterical, yells at her children, who are clambering over the tractor and combine parked in the shed.

"It's O.K., they can play there. They won't get hurt," Mary reassures her, flapping her apron at Gypsy to move, so the children can have easier access to the shed.

Peter Gilmour, one of the chief guides, summed up his observations of the Kosters: "I think the Kosters are more like American Indians than like other Americans, even if they don't realize it," he said. "They don't ever say, 'This land is mine,' or do anything to exploit the people who come here. They simply welcome everybody and say, 'Come share it with us.'"

The Kosters not only donate the use of their land, and tolerate the tourists, but each year, in a small ceremony, they take out a twenty-five-dollar membership in the Northwestern Archaeological Program. They walk out to the table where guides sit selling memberships. Mary writes out a check and hands it to Teed, who solemnly presents it to the guide on duty.

The Kosters have quietly made much more generous gestures for the convenience of the expedition.

When the wooden bridge over the creek began to show signs of wear after a few seasons of tourists' cars rumbling over it, they replaced it with a cement bridge. Teed did the work, with Mary keeping a sharp eye on how it was coming along.

One summer the guides talked, as they sat in the old glider, about fixing up the old Clendenin pioneer rock house to use for slide lectures to show to tourists. When Gordon Will visited the farm the following March to look over the old house, Teed said, "Mary and I cleaned out the old rock house. Thought you might like to use it."

And in 1975, when the "arkies" returned to reopen the site, they discovered that the Kosters had oiled the dirt road leading to their back yard and the cornfield to keep down the dust.

On the Saturday morning of NAP Members' Weekend each year, it has become customary for the Koster crew to receive a special treat as they arrive at the site. A solitary bagpiper, wearing a balmoral tam and red suspenders to hold up his jeans, marches smartly back and forth, piping them onto the site. Bounced against the bluffs at the back of the site, the glorious sounds of "Amazing Grace" and "Scotland the Brave" waft out over the valley. This is one of piper-artist George Armstrong's contributions to NAP's efforts.

Teed's contribution is somewhat less glamorous. On Members' Weekend Saturday there are so many cars at the site that they overflow the back yard and the dirt road. Teed moves livestock out of one part of the pen to make room for more cars. Ever the thoughtful host, he spends the early morning hours shoveling hog droppings out of the pen so that they won't soil his visitors' shoes.

On occasion the Kosters express amusement at their new situation.

One night, after a picnic with some guides, Mary stood up and said, "Come on, Teed. Let's go home to the Koster site."

And they both laughed.

II

Koster,
the Amerindians,
and the New
Archaeology

8

The New Archaeology

When I asked Teed's permission to invite the public to watch our crew as they excavated the Koster site, one of my aims was to educate people about archaeology, as it is practiced today. Many people still retain what I call the "King Tut" view of archaeology. They think of an archaeologist as someone in a pith helmet and khaki shorts supervising a group of native workers in some exotic place like Iraq or Peru. They assume he's looking for buried treasure, which he will bring back and place on exhibit in some museum.

In its earliest years archaeology was treasure-hunting, but modern archaeology is quite different from that and is changing constantly. I have referred several times to the fact that my colleagues in NAP and I are carrying out experiments in the "new" archaeology.

Archaeology began as antiquarianism; people dug in old ruins and collected ancient items out of curiosity. One of the earliest excavators on record was Nabonidus, the last king of Babylon, who lived some twenty-five hundred years ago. When he discovered buried ruins in his kingdom, he had them excavated. He was interested in them for what they could reveal of religious significance.

Over the centuries, many rich or noble individuals emulated Nabonidus and dug because they were fascinated by items created in the past. During the Renaissance people became interested in the arts and architecture of ancient Greece and Rome, and it became very fashionable among the rich to collect antiquities. This spurred people to dig to meet the demands of the market, and much looting of important material took place. Some nations encouraged the treasure-hunting and competed with each other for collections of antiquities for their museums.

Gradually archaeology aroused the interest of scholars and became a serious academic pursuit. In the century or more that archaeology has been a scholarly discipline, it has developed into two main branches, classical archaeology and anthropological archaeology.

Classical archaeology, which usually is taught in the classics or humanities departments of universities, places its emphasis on the study of those parts of the world which contained the beginnings of Western civilization and other early "high" civilizations. (A "high" civilization is a society in which people have developed complex social forms and arts, architecture, and writing. In the New World, the Mayan culture would be considered a "high" civilization; the North American prehistoric cultures would not, because they developed no complex architecture nor any known form of writing.)

Classical archaeologists have traditionally concentrated their efforts on the excavation of public works, and the places where the elite of these early "high" civilizations lived and ruled. Their primary interest has been the study of the history, arts, architecture, and languages of these early civilizations, most of which have been found in the Mediterranean, including Greece, Rome, Egypt, and Iraq.

In the United States, archaeology has developed as a branch of anthropology. Anthropology is the study of the total range of human behavior, past and present, in an attempt to explain similarities and differences in culture, both in time and space. If we assume that all the people on earth today belong to one variety of one species of one genus (*Homo sapiens sapiens*) and, therefore, all of us are the same biological organisms, except for some obvious physical and psychological differences, then how do we account for the vast differences between cultures in Ghana, the state of New York, and Tokyo, Japan? That is the major question to which anthropologists address themselves. Cultural anthropologists study living cultures; archaeological anthropologists study extinct cultures.

There is also a distinction between prehistoric and historic archaeology. Prehistory designates the long time span before human beings developed writing. Humans, in various evolutionary forms leading to modern *Homo sapiens*

sapiens (who evolved, many scholars think, about 35,000 years ago), have been on earth for more than two million years, yet it was not until about 5,000 years ago in the Eastern Mediterranean area that they discovered how to put down abstract symbols which, when examined by other persons, conveyed complex thoughts. Historic times are considered to have begun about 3500 B.C., when the Sumerians developed writing in what is now southern Iraq. Prehistoric archaeology concerns itself with cultures that left no written records. Historic archaeology studies cultures that had developed writing.

Had I been born a hundred years earlier, I would not have been able to become an anthropological archaeologist, studying the prehistory of the North American aborigines. It was not until after 1859 that most people recognized that human beings had a prehistoric past. When the notion was suggested by a few scholars before that, it was largely rejected.

Charles Darwin paved the way for the development of anthropological archaeology when he published his theory of evolution in *The Origin of Species by Means of Natural Selection* in 1859. When Darwin suggested that most species of animals and plants, including human beings, had evolved from much earlier forms, his views produced cries of outrage from theologians and others, and a great deal of controversy arose.

For a long time, fossils had puzzled scholars and theologians. Fossils, they argued, could not represent anything that once had been a living thing, since God created the beasts of the earth on the sixth day, after He created the dry land. Consequently, there was no way that animals or fish might have become embedded in rocks on dry land, went the argument. Some thought that fossils were the work of fairies or of the Devil.

At the time, even educated people accepted as true the theories of James Ussher, Archbishop of Armagh and Primate of the Church of Ireland, who in 1650 had worked out a chronology of the history of the world, based on lengths of time mentioned in the Bible, setting the creation at 4004 B.C. His dates were inserted as marginal notes in many editions of the Authorized, or King James, Version of the Bible, giving him so much authority that when Sir Charles Lyell and other early geologists came out with opposing theories, they were met with firm religious opposition.

In 1830 Lyell published *The Principles of Geology*, in which, borrowing from the ideas of Dr. James Hutton, a Scottish physician with a passion for examining ideas, he set forth the principle of uniformitarianism. This principle states that the natural processes affecting the surface of the earth—including landforms, water resources, mountains, etc.—which occur today also occurred in the past. Therefore processes in the past can be interpreted in terms of the processes that can be observed at work today. Using this principle, Lyell demonstrated that the forces of wind and weather had slowly altered the shape of the earth's crust over millions of years. If one accepted Lyell's theories, one could explain fossils. Since some of the fossils resembled skeletons of creatures still living, one could assume that previous animals or fish had existed and had died, leaving behind skeletons which became embedded in rock faces.

At the time that Darwin published his theory of evolution, there were no known fossil human bones except for a skull that later came to be known as Neanderthal man, discovered in 1856 in a cave near Düsseldorf, Germany. There were great arguments raging over whether this was the skull of a primitive form of human being or of a contemporary idiot.

As Darwin's book was being published, Jacques Boucher de Crèvecoeur de Perthes, a French scientist, was having trouble getting other scientists to accept his theories that rude flaked-stone implements which he had discovered near Abbeville, France, were hand axes made by people in Paleolithic times. This would have made the makers of the axes contemporaries of extinct animals. Unfortunately for Boucher, his workmen, observing their leader's enthusiasm when he found these rough tools, thoughtfully manufactured some flint instruments and buried them where he was digging. These were revealed to be forgeries; all of his finds became suspect; and he was scorned by his fellow scientists.

In 1859 evidence was found on the floor of Brixham cave in England that early people had lived there. Flint tools were found along with the bones of lions, reindeer, bears, mammoths, and rhinoceroses, all of which were by then extinct in England. To assure that there was no fraud involved, a distinguished committee of scientists was appointed to supervise the excavation, and even skeptics were convinced. Later these scientists visited Boucher and confirmed his finds.

When Darwin published his theory of evolution, he made reasonable explanation of the fossils and tools possible by showing that various species had existed for much longer than had been assumed and that there was evidence for a series of developmental stages in these creatures, going from simple to complex form. Darwin not only had an enormous impact on the field of biology but also influenced the study of the history of human cultures on earth.

Anthropology as a scholarly discipline was then in its formative stages. Lewis Henry Morgan, one of the great early American anthropologists, after studying Darwin's concept, suggested that cultures, too, have gone through a set of stages from simple to complex. In his book *Ancient Society*, published in 1877, Morgan outlined a scheme purporting to show that all societies go through two stages of savagery, two of barbarism, and finally two of civilization, in their progress from simple to complex forms. The highest level, in his opinion, was the stage of civilization as exemplified by Western cultures in Europe and America.

There was a great deal of intellectual furor over Morgan's ideas (which no longer are accepted by anthropologists), and this led to a new outlook on archaeology. Eventually, from the concepts stemming from interest in Morgan's theories, anthropological archaeology was developed. Morgan maintained that the stages of cultural evolution he set forth were universal stages of growth for all cultures. Some cultures, he said, had frozen at different levels, which he felt accounted for the vast differences between cultures. Those who wanted to test this hypothesis realized they would have to dig in sites all over the world, and in many places where high civilization had not extended. Thus, anthropological archaeology got its start.

In order to understand the differences between classical archaeology and anthropological archaeology more clearly, it helps to observe how archaeologists of each of these schools treat an individual site, how they regard all the sites within a given region, and how they regard sites on a worldwide basis.

At the individual site level, classical archaeologists dig only selected parts of a site. They are interested primarily in the public precincts of a site, in the palaces, temples, and tombs, and in the great religious and ceremonial centers where members of the culture participated in social, religious, and political activities. They seldom dig those parts of a site that reflect other aspects of the human behavior that took place there.

Within a region, they dig only those sites that give evidence of having preserved architecture, or language texts, or objects of art. They ignore a wide range of other types of sites that may exist in a region.

And on a worldwide basis, classical archaeologists work only in those parts of the globe where cultures have evolved to the stage of high civilization.

Since anthropological archaeologists see as one of their goals the study of the full range of human behavior, at an individual site they think it is necessary to analyze all areas where human activities took place and not just the palaces or temples or tombs. They dig in areas of a site belonging to the most menial aspects of life, such as the houses of the lowest-ranking people in the society, their work areas, their cooking fires, and garbage dumps.

On a regional basis, anthropological archaeologists study all the sites that existed within the area, including the most minor hamlets, hunting camps, butchering sites, and cemeteries of people of all ranks.

And, finally, anthropological archaeologists dig anywhere in the world where they think a site may yield information about differences in cultural development.

From now on, when I refer to archaeologists, I will generally be talking about *anthropological* archaeologists.

Every once in a while there appears in the press an item about how archaeologists are using modern technology to help locate a site or to analyze data. In reality, archaeologists lag far behind the rest of our society in applying twentieth-century technology to their work and tend to be resistant to new methods and ideas.

In the mid-1930s a radical new idea was put forth by archaeologist Waldo Wedel. He suggested that in addition to studying artifacts and other material remains left by ancient people, archaeologists should examine the interrelationships of plants, animals, and human beings as they had existed together in the past. His colleagues paid little attention to his suggestions.

By 1948 archaeologist Walter Taylor was more directly

critical of his colleagues. Archaeology, charged Taylor, should be approached as a form of anthropology, but it was not. He pointed out that as it was then (and for the most part, still is) being practiced, its main goal was to set up a timetable of excavated artifact groupings. Most archaeologists spent the major portion of their energies in describing how the artifacts they had dug up looked and how old these artifacts were in relation to other artifact groupings from other sites. Then they would define an extinct culture in terms of the artifact assemblages that were its only surviving remnants.

This, insisted Taylor, was not describing a culture at all; it was simply describing a series of standardized artifact forms preserved from an extinct culture. The problem that had not been adequately addressed, he said, was how to use the artifacts to arrive at an understanding of the cultures that made them. It is the customary forms of behavior, which are only in part reflected by the artifacts, that archaeologists should seek to understand. While many archaeologists attacked Taylor, others found his ideas provocative and began to discuss them and to look for ways to put them into practice.

In 1952 Grahame Clark, Disney Professor of Archaeology at Cambridge University in England, in his book *Prehistoric Europe: The Economic Basis*, expanded the ideas on which Wedel had touched briefly. Clark suggested it might be appropriate to apply the concepts of ecology to the study of prehistoric people. Like Wedel, he pointed out that human beings are only one element in the natural world, and that to understand how and why prehistoric people shaped their cultures as they did, archaeologists should study the natural surroundings, including animal and plant life, as they were when prehistoric people lived among them. Archaeologists, said Clark, should attempt to understand how ancient people developed strategies to use the specific characteristics of their environment effectively. He felt they should look at how people had coped with aspects of the environment that were difficult and dangerous, and how they had exploited others that had the potential to make life easier and more secure.

From 1948 on Robert J. Braidwood, of the University of Chicago, was making plans to excavate a site in Iraq. Fascinated by Clark's ideas, he took along a geologist, a botanist, a zoologist, and a ceramic technologist. Each was assigned to apply his or her specialty to figure out what the plant and animal resources in the area had been in prehistoric times, the strategies these ancient people had used to procure and use certain of these resources efficiently, and, finally, the process by which certain wild plant and animal species had become domesticated, since domestication would have had great impact on the basic economic lives of these people. His was among the first efforts to develop a multidisciplinary team of specialists capable of studying the central problems of human beings' relationships to their environment.

Braidwood's pioneering efforts were followed by the work of Richard MacNeish. In 1948 MacNeish, then at the University of Chicago, set out to trace the origins of corn cultivation in the New World. Simultaneously he studied the cultural changes which accompanied this important development. Eventually, in 1960, he found one of the places where corn first was cultivated (circa 7000–5000 B.C.), in the Tehuacán Valley, a semiarid area in the southern highlands of Mexico about eighty miles south of Mexico City. MacNeish organized what was then by far the most complex interdisciplinary team ever used in archaeology.

Starting in the mid-1950s, there arose a great deal of controversy among North American archaeologists. One small but vocal group who accepted the views of Wedel, Taylor, and Clark became very critical of the theories and methods of archaeology and of the accepted ways of performing it. They became the new archaeologists.

One of the fundamental differences between traditional archaeology and new archaeology is the way each defines culture.

Traditional archaeologists define culture as a body of ideas, values, and belief shared by a group of people. One generation passes these ideas, values, and beliefs on to the next. Archaeologists look at the material remains of an extinct culture in order to define what concepts were in the minds of ancient people.

The new archaeologists define culture as a nonbiological system which human beings develop to cope with their environment. Culture is seen as a series of interlinked behavior patterns and material items. All parts of the cultural system—economic, social, political, religious—are interdependent. The total cultural system serves the people who developed it as a buffer between them and their environment. It permits them to shape the environment to their ends, to protect themselves against danger, and to exploit potentially valuable aspects of the environment effectively, including not only animals and plants but other cultures as well.

In setting their goals, traditional archaeologists have been influenced by psychology; they are asking, What were the mental modes of the people who made these artifacts?

Instead, new archaeologists employ a set of concepts derived from evolution and ecology. Cultural systems go through a set of changes, becoming increasingly complex. As a system becomes more complex, it develops more parts, with each part performing fewer tasks but performing these more efficiently. Therefore, the total cultural system enables human beings to meet their needs better in their environment. New archaeologists ask, What are the reasons for these changes? When, why, and how do cultures become more complex? How do human beings interact with their environment as the changes occur?

Although archaeologists of both schools aim to describe and explain prehistoric cultures, they differ in what they mean by "description" and "explanation."

Traditional archaeologists excavate sites in order to find the formal representations of the ideas that ancient people shared, as seen in their artifacts and other material remains. They feel that by examining material remains and making up a trait list of the standard forms of houses, spearheads, knives, pots, and burials (among other artifacts and features) they can infer what basic ideas, values, and beliefs these people once shared. They would then use these inferred ideas to *describe* an ancient culture.

The new archaeologists also study artifacts and features, but they want to use these to understand how the cultural system was used by ancient people to cope with the environment in which the culture existed. Therefore, they study the extinct environment too. And some new archaeologists have added another vital element to their studies—the remains of human beings who once lived at a site. (Until recently, archaeologists have studied human remains from North American sites in a very limited way.)

Thus, when new archaeologists try to figure out how well an extinct group adapted to its environment, they try to reconstruct on paper three extinct systems—the environmental system, the human biological system, and the cultural system. By figuring out how each of these three systems interacted with the others in a given culture, the archaeologists then can explain changes in any one, or all, of these systems.

As they excavate, new archaeologists collect various kinds of material that traditional archaeologists have ignored.

All of us, as we move through the day, create various kinds of debris, much of it unnoticed. We drop bits of food, gum wrappers, paper clips, pop-can tabs, a piece of broken shoelace; the list could go on and on. New archaeologists call this debris the by-products of everyday maintenance, and they collect it. Traditionally it has been tossed out by archaeologists. On a prehistoric site this might include such things as tiny chert or flint chips, pieces of broken tools, bits of animal bones, and other discarded materials. New archaeologists also collect bits of evidence that might reveal what the prehistoric environment was like, such as pollen, snail shells, and fish scales.

When Tom Cook, who analyzes stone artifacts from Koster, examined the material from Horizon 6, he carefully analyzed the debris that Helton people (whom he had named for Alec Helton) had dropped as they went about their everyday lives. He was looking for evidence of activity sets, or areas of the site where debris can be found to indicate what specific activities had taken place at that spot.

Then Tom was able to prove, for example, that the Helton people had been manufacturing certain tools and repairing others while living at Koster.

The traditional archaeologists would have arrived at their *description* by describing the artifacts, and then *inferring* activities from them. Tom arrived at his *description* by examining the behavior of the Helton people from the activity sets. And he only stated conclusions about their activity (or behavior) that he could *prove* took place.

Another difference involves the definition of *explanation*.

For example, on the high plains of North America for thousands of years prehistoric Indians hunted buffalo on foot. In the 1500s the Spaniards arrived in the Southwest, bringing guns and horses. These cultural items, guns and horses, were adopted by the Indians, who had observed how they could affect the way buffalo were taken. The Indians thereupon developed new forms of culture which included the techniques of surround hunting rather than the old style of ambush. Tremendous social, political, and religious changes occurred among the tribes as they changed from hunting on foot with bows and arrows to hunting from horseback with guns.

Traditional archaeologists would explain this by saying that what happened was the diffusion of two cultural traits (horses and guns) from Western European culture into Great Plains cultures.

New archaeologists would say that that is no explanation at all. They would ask why the Great Plains cultures took over only certain traits from the white culture and not all. They would offer an explanation in terms of the shifting demands of the environment in which the Great Plains cultures existed. The Indians added horses and guns to their technology because these gave them a greater chance of survival. They could hunt much more efficiently and also better defend themselves against a new threat to their survival, the Europeans. Other traits from Western European culture did not appear to offer chances for greater survival and were rejected.

Traditional archaeologists also explain cultural change by independent invention. About A.D. 500 prehistoric people in the American West began to build earth lodges, which were much more substantial than the flimsy structures they had previously constructed. The usual explanation is that this shift occurred because someone invented the earth lodge.

The new archaeologist would try to discover what had been taking place in the social or physical environments of these people at the time they began building earth lodges. Perhaps the weather changed, bringing longer cold seasons, so they had to build sturdier shelters to cope with them. Perhaps they began to cultivate crops and as a result became more sedentary. These theories could be considered and tested to explain this cultural change.

New archaeologists maintain that the past is knowable and that the accuracy of knowledge about the past can be measured if the data are subjected to the scientific method. They insist that archaeological data, like data in the physical and biologial sciences, should be subjected to rigorous scientific analysis.

It is not enough, say the new archaeologists, to look at a scattering of chert chips on the ground, with a deer antler lying in the midst of them, and say, "This looks like a spot where some worker was repairing stone tools." They insist that one must *prove* that someone was repairing tools there.

In using the scientific method to test data, the scientist examines the evidence and derives a hypothesis or theory about it which he or she thinks may explain all examples of a phenomenon, then uses the data to test this hypothesis, in a logical sequence. From the results certain conclusions are drawn. To be acceptable to other scientists, these conclusions must be capable of being reproduced by independent testing of the data. If the test shows that the scientist's hypothesis was incorrect, he or she formulates a second hypothesis and proceeds to test it, repeating the process until a provable theory is found.

One method new archaeologists use to reconstruct the behavior of extinct people is to study the subsistence settlement systems of prehistoric cultures, or the group's game plan for getting along in its environment.

All of us have game plans for getting along in our environment, although we may not be aware of them as such. We set aside certain portions of the day and week for certain activities, and perform those activities in special places. Some hours of the day are for eating, dressing, working, shopping, recreation, sleeping, worshiping. These activities take place in our homes, offices, shops, tennis courts, bowling alleys, churches.

For prehistoric people at Koster, the environment consisted of prairies, forests, floodplains, bluff tops, rivers, lakes, and streams. Just like modern people, they had to plan where they were going to invest their time, to eat, make tools, hunt, collect wild foods, have recreation, or sleep. We want to find out what their strategy was for obtaining energy, in the form of animals and plants, from their environment.

Like any good detectives, archaeologists start with the smallest clues and keep trying to add to these, or to match clues, to reach a provable conclusion. For example, the combination of fish scales found with a stone scraper, after being tested scientifically, might reveal that someone once cleaned fish at the spot; this would be labeled an activity set. Two or more activity sets comprise a settlement type. The Helton village at Koster Horizon 6 was a settlement type. Two or more settlement types constitute a settlement system.

While Helton people at Koster Horizon 6 might have lived at Koster all year round, at different times of the year small groups might have gone out and lived elsewhere temporarily to gather necessary foods or raw materials. They would have gone out on hunting or gathering trips according to the season.

In the fall some of the women, taking infants and small children with them, might have moved down to the river floodplains for a while to collect the seeds of wild plants such as marsh elder (*Iva annua*) or pigweed (*Amaranthus.*) Since their aim would have been to gather enough seeds to see their families through the cold winter months, they might have stayed there, camping, for several weeks, until they had collected the seed crops in the area. Their temporary camp would be considered a settlement type; and as we attempt to reconstruct their settlement system, we shall look for potential archaeological sites in the floodplains.

During the winter some of the Helton men might have moved away for two or three weeks, east into the secondary valleys or out onto the prairies, to set up camps to hunt American elk and white-tailed deer. They would have brought carcasses back to the hunting camp, where they would skin and butcher the meat, some of which they would dry before carrying it back to the village at Koster. This would be another settlement type.

All three—the village at Koster, the seed-processing camp, and the hunting camp—would be part of the subsistence settlement system of the Helton people.

Since we want to locate the various settlement types that might have made up different settlement systems in prehistoric Lowilva, we have to look at many sites. We have staked out, as our "research universe," a 3,200-square-mile area of Lowilva—an area large enough to include all the potential resources that might have been exploited by any of the cultures we want to study over a time span from 10,000 B.C. to A.D. 1673.

Each summer our NAP site-survey teams, under the direction of Kenneth Farnsworth, fans out over this area looking for potential sites. They map sites, collect artifacts from the surface, and place their notes in the computer site-survey file. With a considerable amount of help from our thirty-five local artifact-collectors, the site-survey team has mapped more than twelve hundred prehistoric sites in the region.

After we have excavated Koster and analyzed the data taken from there, we will try to figure out which, if any, of the twelve hundred prehistoric sites in Lowilva might also have been occupied by any of the Koster groups. As we succeed in this, we should build a rich and accurate picture of prehistoric Indian life at Koster, and in Lowilva.

9

What Was the Climate Like in 7500 B.C.?

The climate of Inuvik, up near the Arctic Circle in Canada's Northwest Territories, is in sharp contrast to that of Lagos, Nigeria, on the west coast of Africa. There is little green foliage in Inuvik except for some short, scrubby growth during the brief summer season. In coastal Lagos the tropical growth is thick and luxuriant.

While an Eskimo at Inuvik and a Nigerian may both work outdoors, the fur-lined parka and pants donned by the Eskimo in the Arctic winters would drive the Nigerian mad in moments if he wore it in his homeland. Until modern North America cultures impinged on the Eskimo's, he lived in winter in a home made of ice which provided excellent insulation from the cold. The Nigerian may live in a house made of very lightweight materials—upright posts and a thatched roof designed to shield him from the searing tropical sun. Until the introduction of other foods by Euro-Canadians, the Eskimo lived mainly on seal and caribou meat; the Nigerian eats fresh fruits and vegetables all year round.

The physical environment in which any group lives plays a vital role in determining how it will live. If we wish to learn how Koster people got along in their natural surroundings, we must first try to determine what the environment was like when they lived in Lowilva.

Much of the work we are doing with archaeological remains is still in the experimental stage. We are trying to reconstruct on paper an extremely complicated ecosystem.

What kinds of changes have taken place in Lowilva in the last ten thousand years—in the rivers, creeks, lakes, swamps, floodplains, bluffs, hillside slopes, and upland prairies? What was the climate like during that period? We must try to determine how hot or cold it was, how much rain fell, what kinds of winds prevailed during all that time. In addition, we want to know what kinds of plants grew there, and what kinds of animals were able to thrive.

The NAP botany laboratory is housed in a small white house on a side street in Kampsville, across from St. Anselm's Church parish hall. The director of the botany laboratory is Nancy Asch, a paleoethnobotanist (a botanist who studies ancient cultures). Her husband, David, an archaeologist and NAP's statistician, works with her, analyzing plant remains.

In one corner of the laboratory sits a vase filled with dried cattail stalks. On Nancy's desk, face down, lies a copy of *Earth Basketry* by Osma Gallinger Tod. Nancy plans to try her hand at weaving baskets with material gathered from the fields near Koster, just as Amerindian women did thousands of years ago.

Nancy is trying to get a picture of what Lowilva looked like when the first Euro-Americans arrived to settle there in about 1820. It is difficult to project back from what one sees today in Lowilva. The land surface has been radically altered by human activities; the floodplains have been drained for agriculture, and there are levees built along the Illinois River to prevent flooding.

Nancy consults records of the U. S. Government Land Survey to get a picture of Lowilva before the first Euro-Americans arrived. From 1815 to 1820 surveyors marked off the land in square sections, one mile on a side, and made observations on vegetation within every quarter section over the entire landscape. They identified "witness" trees and noted whether the land was covered with forests, prairies, or brush.

The next step is to look at the plant remains from Koster. To aid her in identifying charred prehistoric plant remains, Nancy spends time in the fields and forests of Lowilva, hunting modern plants for a reference collection.

She plans to collect, identify, and mount a specimen of every plant and tree that grows in the 3,200-square-mile area designated by NAP as its research province. She has marked the area off on a large map, and systematically collects from different sections. For each specimen, she gathers an entire plant, including roots, stems, leaves, flowers, and seeds. In the case of trees, she collects leaves, flowers, fruits, sections of wood from trunk and branches and a piece of root. She goes back to the same stand of plants or trees in different seasons to collect specimens at various stages of growth.

At the botany laboratory Nancy prepares the modern specimens. Each is identified, pressed, dried, and mounted on a large sheet of paper. These are stored in an airtight cabinet with mothballs.

On Nancy's table is a small box filled with sassafras seeds. She takes one, splits it neatly in half with a sharp knife, then glues one half, cut side up, to a glass microscope slide. Next she glues a whole seed to the slide. She is building a collection of seeds and bits of wood from all species of trees and shrubs in the region. The seed collection would enhance any art gallery; the arrangement of the seeds and the colors of the seed coats and their interiors make each slide a small work of art.

Nancy burns some of the modern samples enough to char them, so that they will more closely resemble the archaeological specimens. Ancient plant remains found at a site have survived only in charred form, because they were dropped or fell into the fire and were incompletely burned. The rest of the plant remains have long since decomposed. Nancy drops fresh plant specimens into sand that has been heated in a Dutch oven on the stove, or she may place some in an airtight crucible which she heats over a Bunsen burner. She keeps the specimen in the hot sand or over the burner until it stops smoking. Then she mounts sample bits of the charred fragments on glass slides.

Kathy Freudenrich, one of Nancy's research assistants, sits before a microscope at a worktable. She empties a small pile of archaeological remains onto a petri dish or sorting tray. Using a fine artist's brush, she gently separates them under the microscope. If she can identify a charcoal fragment, Kathy puts it into a small labeled box. If she cannot identify it, she turns to the reference collection of slides containing charred plant specimens and compares the archaeological specimen to them until she can tell exactly which species of plant it is. After the samples have been identified, they are counted, weighed, measured, and stored in gelatin capsules or plastic bags. All the information is then placed on computer coding forms for use in later analysis.

Identifying plant remains takes a sharp eye and a great deal of patience. Between 1970 and 1975 Nancy and her crew had identified more than a hundred thousand charred bits of nuts, seeds, and wood.

Usually a site holds so much material that it would produce massive quantities of redundant data if all of it were dug up. To avoid this, archaeologists use sampling techniques and dig up only portions of a site. They take samples of material from all areas of a site where human activities are believed to have occurred. It is important that they collect samples of all the human activities which took place, in sufficient quantities to assure an accurate assessment of the data. The degree of efficiency with which a site is dug depends on the accuracy of the sampling technique. David Asch, who is an experienced statistician as well as an archaeologist, uses modern statistical methods to help NAP decide where to take samples on the sites we dig.

We use another sampling technique, called flotation, to collect archaeological remains which are not caught on the screens at a site. Have you ever watched dead leaves floating on the surface of a pond or a puddle? The principle behind that phenomenon, namely that dead plant matter is lighter than water, helped me solve the problem of how to retrieve some very important material from archaeological sites—tiny charred bits of seeds and nutshells, and pieces of animal remains. Many of these are so small that they remain invisible to the naked eye even if you crumble the soil in which they have been embedded for centuries.

Before I developed a method for retrieving this material, along with most other archaeologists, I had been throwing it out, largely unaware that it could be collected. It was an observant botanist who first suggested to me that there might be a way to save these important clues.

In 1959 as we were excavating Kamp Mound Number 9, I noticed that in some of the earth in our trowels there appeared to be small bits of charred materials. I examined these and decided they were fragments of seeds. I was much interested. One of the most important links between a human population and its environment is, of course, the food it eats. I was curious to see what this evidence could tell us about the Hopewellians' diet and their food-gathering habits.

Later, as I looked through the items collected from the screening process at the site, I realized that the bits of charred seeds, falling through the screens, were being tossed out with the dirt.

As we dug, we came across a pit that contained an extremely ashy fill. I examined it and could again see tiny bits of charred seeds. I put aside a fairly good-sized sample until I could figure out a way to remove the seeds from the dirt.

At that time I had a visitor, Dr. Hugh C. Cutler of the Missouri Botanical Garden in St. Louis. Hugh is an expert on early domesticated plants in North America. He is interested in tracing the beginnings of agriculture in the New World and in trying to discover where, how, and when corn, squash, pumpkin, and other plants were first domesticated in what is now the United States. Hugh was especially interested in what we were finding at Mound Number 9, because agriculture is supposed to have begun in the North American Midwest during Hopewell times.

Hugh examined the sample I had saved from the pit at Mound Number 9 and suggested that the burned bits of plant material would float to the surface if the soil was put in water.

So I took a sample of soil from the pit, placed it in a bucket of water, and stirred it. Little black objects began to float to the surface, and as they appeared, I scooped them off with a tea strainer. When I shook out the contents of the tea strainer on some newspaper, I was amazed at the quantity of tiny pieces of charred seeds I had gathered, since many of these had been totally invisible when they were encased in soil.

Over the next couple of years, I kept experimenting with the flotation technique, trying to improve it. The process was not yet refined enough for our purposes. When a sample of dirt from a site is placed in water, flotation yields two sets of products. One, called the light fraction, consists of bits of residue from people's daily meals and includes pieces of animal bones and carbonized plant remains. The heavier fraction contains bits of stone, burnt clay, small pottery fragments, and various other artifacts, such as an occasional bead or tiny stone drills. The technique was not wholly satisfactory because the light fraction contains both carbonized plant remains and bits of animal bones. Each of these types of materials must be sent to a separate laboratory for analysis. They must be separated, and to do this by hand, using a tweezer, as we were doing, takes a great deal of time.

Help came again from an unexpected quarter. Two of my students were Sue Bucklin and Joanne Dombrowski, who were also students of Laurence Nobles, professor of geology at Northwestern University. Sue and Joanne had shared in our frustration when they tried to hand-separate plant and animal remains with a tweezers. Back on campus, they mentioned our experiments in flotation to Nobles. He described a chemical flotation process used in the coal mining industry employing a zinc chloride solution of a specific gravity which allows the coal to float off and the residue to sink. He suggested we try using a similar solution to float off the carbonized plant remains.

I provided Sue and Joanne with samples of the light fraction, which they took to Evanston. In Nobles' laboratory, they tried the chemical solution and were able to achieve one hundred per cent separation of plant remains from animal bone remains.

Today NAP runs samples of soil from all the sites it excavates through both water and chemical flotation processes. Each day a crew picks up the half-bushel baskets of flotation samples caught under our table screens at the site and takes them to the Illinois River for the first step in the process.

Passengers using the Kampsville ferry frequently are treated to an impromptu sideshow during the excavation season when the NAP flotation crews take to the river. To float a sample, a worker stands hip-deep in the river holding a galvanized tin washtub. The bottom of the tub has been removed and replaced with one-sixteenth-inch fine mesh screen. The person holding the tub swishes it back and forth in the water, often wiggling his or her hips from side to side in accompaniment. The action sometimes resembles a belly dancer's gyrations. As the tub is gently rotated, the fine-grained silts and sands wash out. Within a few moments only a residue remains. This may include flint chips, animal bone fragments, fish scales, pieces of stone artifacts, and charred seed and nut remains.

Each tub-wielder has a partner, who works with a large hand strainer composed of a metal frame and the exceedingly fine brass screen mesh used in automobile carburetors. We have these made especially for us in a machine shop in nearby Carrollton, Illinois. When the residue from the first tub-shaking settles into the bottom of the tub, the first person dips the tub into and out of the water very quickly, so that the various materials are lifted by the water and then settle again at different rates. The heavier fraction, such as potsherds and pieces of chert, settles fastest. The lighter fraction, which contains animal bones and plant remains, settles more slowly. The person holding the strainer must quickly skim off the lighter fraction before it sinks.

Together the teams of workers dip the tub and strain out the light fraction over and over. The material removed in the strainers is laid on absorbent paper to dry. Although the procedure is simple, each two-person team must learn to work together, co-ordinating tub and strainer action.

After the light fraction has been dried, it is put through the second step of the flotation process in the laboratory. The workers make a bath of zinc chloride solution, and then pour a small amount of the light fraction into the bath. Animal bones go to the bottom; snail shells and carbonized plant remains float to the surface. Fish scales vary: some float, some sink to the bottom.

When we began to excavate Koster in 1969, I expanded the flotation activities. We began to collect more samples of soil for flotation from Koster than we had at any previous sites. And Nancy has told me that the flotation process is producing more than ten times the amount of plant remains and charcoal from Koster and other sites than we had been obtaining by using only the screening method.

These larger quantities of charred plant remains, and the increased number of contexts in which charcoal and plant remains are found, have enabled Nancy to estimate the prehistoric people's plant-food uses with greater accuracy. This is especially important at Koster, because where there are changes in the use of wild plants from one occupation level to another, there appear to be differences in emphasis on certain plant foods rather than complete shifts from one type to another.

Sometimes people ask the Asches how they can make reliable reconstructions of prehistoric environments from charred archaeological samples, given human beings' highly individualistic tastes in foods, and because only certain parts of plants are preserved by chance. As scholars, the Asches use a very conservative approach. They bypass the question of personal choice altogether, and draw implications about a species of plant only if it shows up in the archaeological remains.

We can tell a lot from the presence of a species of plant or tree. If you know the ecological requirements of that species, its presence implies certain environmental limits. For example, the northern distribution of pecans is controlled by the length of the growing season. Remnants of pecan shells have been found in all of the horizons at Koster, and that lets us know that while people lived there, the growing season for plants in Lowilva apparently was not much different from what it is today, since pecan trees are still found in the region.

Cattails grow best in moist places; cacti thrive in the desert. Because different species of plants and trees prefer to grow in specific habitats, the presence of a particular species in the archaeological specimens can tell us what kinds of plant habitats existed in Lowilva in prehistoric times. For example, residents of every village at Koster apparently liked black walnuts, for we find shells from these in every level. This lets us know that there must have been moist forests near the site all during its various occupations. We also found seeds of marsh elder in the debris from every village and hunting camp, indicating that during all of the Koster occupations there must have been areas of open, wet, marshy ground in Lowilva.

Between 5000 and 2000 B.C. (when there were large settlements at Horizons 8 and 6 and a much smaller hunting camp at Horizon 4) the Koster people lived in a very stable vegetational environment. In all of those settlements, we found the same wood charcoal, nut, and seed specimens, although in varying amounts.

Apparently the climate in Lowilva had changed very little over the centuries. What change took place was not pronounced enough to cause the disappearance of any one kind of plant in the region and its replacement with another. But the degree of climatic change did affect the size of some plant communities, causing some to expand and others to contract. These changes would have had significant consequences for the Koster people, for they would have affected the availability of various plants for food. We think the bluffs which line the valley on both sides of

the Illinois River served as a buffer for the effects of
regional climatic changes on the plants and trees within
the valley.

When Nancy compared the early U. S. Government Surveyors' records with the Koster data, she noted that the
plants, trees, and shrubs which had existed in Lowilva
when the first Euro-American settlers arrived were strik-
ingly similar to those which had been there during all of
the Koster occupations. Furthermore, the plants and trees
which had flourished during the whole time that Koster
was occupied are the same as those found in Lowilva today.

Nancy's work with charred plant remains is complement-
ed by the work of NAP palynologists (pollen specialists),
who also reconstruct the prehistoric environment, by comparing modern pollen samples to ancient ones retrieved
from the site.

Next time you are in the garden, stick a finger into a
flower and pick up a few grains of pollen. They may look
fragile, and if you blow on them, some will float away, but
those tiny pollen grains are among the most durable items
nature makes. Thousands of years from now, if archaeologists dig up your garden, some palynologist may be able to
tell whether you grew flowers for color or fragrance or
whether you helped feed your family by growing vegetables.

Pollen has a unique biochemistry that makes it very
resistant to just about everything but oxidation and mi-
crobes. If left exposed to air, it can be destroyed by micro-
organisms; but it is very resistant to normal soil chemicals,
and once it is buried, it will last indefinitely in certain soil
conditions. Acidic bog deposits are an ideal source for fossil pollen, but pollen also may be found in silts and clays.
At Koster the fossil pollen was preserved in redeposited
loess.

Pollen produced by flowering plants is set free to float
in the air, and if it happens to drop into the proper soil it
will be preserved. Pollen from different species can be
identified, and the palynologist can figure out what plant
communities grew at a site in prehistoric times by studying
the pollen specimens found there. Again, since climate af-
fects vegetation, when the plant (and pollen) pattern in a

site or region changes, the palynologist can discern changes in the area's weather.

Dr. James Schoenwetter, a palynologist at Arizona State University, and Rose Duffield, director of the NAP pollen laboratory, are reconstructing the prehistoric plant communities at Koster. If vegetation patterns changed, were these shifts the result of climatic or human agencies? When people settle in a place, they usually change the landscape, and, in turn, affect the plant and pollen record. Sometimes they clear land to live on; they cut down forests for wood for houses or fuel. They may plant crops. By comparing the types of pollen which accumulated while the Koster site was occupied with the types of pollen which fell when the site was abandoned, the palynologists can determine what impact, if any, the Koster residents had on their habitat.

Human impact would show up as disturbances of the natural environment, encouraging the growth of weedy herbs and shrubs more characteristic of disturbed or edge situations. This can be seen very readily in the pollen record, because many of the weedy species produce vast amounts of pollen.

Rose's object is to present a picture of the vegetation makeup of Lowilva (similar to that which Nancy is putting together) and of past climatic shifts. As we study the settlement patterns of the people who lived in Lowilva before the Euro-Americans arrived, we need to understand the vegetation makeup of each of the various zones in the region—floodplains, talus slopes (the bottom half of hillsides), bluff crests, and rolling uplands—through time. Because of the sharp physiographic differences in these zones, they supported diverse plant communities and their potential for human beings was varied. We cannot take it for granted that the vegetation in each of these zones was identical during the different prehistoric periods when people lived in Lowilva.

The palynologists collect modern samples of pollen to be used as a reference bank for identifying prehistoric specimens. The collection also serves to document the distribution of modern pollen in the region.

For the reference collection, the palynologists collect two or three plants of a single species, which go into a plastic bag in the field for later identification and pressing.

They also collect a number of blossoms, which are immediately sealed in envelopes.

In the field, Rose also collects samples of soil from just below the leaf layer on the ground. And she collects tufts of moss, which are natural pollen traps.

Archaeological pollen samples are collected in soil samples from every three-inch level dug at the site.

The palynologists also have cores of soil taken from deep in the ground. For this, they use a core-drilling machine mounted on the back of a truck. The machine pokes a hollow tube into the ground and collects about a four-foot length of soil, two or more inches in diameter. When the palynologist examines these core samples, he or she also looks for natural carbonized material in the soil which can be used to date the layer from which the pollen is taken. The coring technique provides one of the best ways to avoid possible contamination of the pollen record. So far, we have cores taken to about thirty-five-foot depths from Koster and other sites in Lowilva and from many vegetation zones in the region.

Core-drilling machinery is very expensive, and we cannot afford to purchase any, but several agencies have obtained core samples for us. The Soil Conservation Service of the U. S. Department of Agriculture, the University of Illinois Department of Agronomy, and the Illinois State Geological Survey all have taken cores for us while studying modern vegetation.

To extract fossil pollen from soil, the palynologist literally dissolves the soil from around the ancient pollen grains with strong acids. The pollen grains, being tough, remain. To obtain pollen from fresh flowers, the palynologist grinds the flowers into a fine paste and then treats the paste in several steps with a variety of chemical solutions.

The pollen grains are placed in a glass vial of silicone oil, which will preserve them indefinitely. Reference slides can be made from this solution, which are then identified and counted under a microscope, by being magnified about four hundred times. The palynologist counts each grain of pollen. The percentages of different pollen types in a count reflect a number of things about the environment which affected the plants that produced the pollens. The palynologist takes these percentages into account when trying to reconstruct prehistoric plant communities and past climatic changes.

Jim Schoenwetter's analysis of the pollen record at Koster suggests that when there was a tiny hamlet at Horizon 11 (6500 B.C.), although the site itself was in an open, dry area, it may have been surrounded by forests. These conditions persisted over several thousand years, on through the time when a large village was occupied at Horizon 6 (3900–2800 B.C.). The pollen spectra from the time when Horizon 9 was occupied (about 5800 B.C.) indicates that the area had become more moist. Pollen from cottonwood trees appears, and they prefer a moist habitat. Presumably these grew alongside the Koster creek.

By the time hunters spent time butchering deer at Horizon 4 (2000 B.C.), the dominant element in the region was moist forests.

Overall, the climate changed gradually from relatively dry at about 6500 B.C. to a more moist stage at about 2000 B.C.

Many ancient plant communities have been destroyed not by environmental change but by development of land for agriculture, roads, and houses. Someday we may find a remnant of floodplain prairie that has not been plowed up, from which we can take a pollen sample. A few years back, before I got so involved with Koster, I used to give speeches before the Farmers' Grange or American Farm Bureau groups, and I would always ask: Does anybody know where we might find a bit of floodplain prairie still intact? But no one ever knew of a square foot of pristine, undisturbed floodplain in the entire seventy linear miles of the Lower Illinois River Valley. Now we figure the only way we will get that particular pollen complex is to try to find a place where a floodplain prairie existed, over which early settlers built a levee. If we could find one, we could cut down the levee to find the original ground surface, and take pollen samples.

In another Kampsville laboratory, NAP archaeologist Irwin Rovner, assistant professor of anthropology at North Carolina State University, is trying to reconstruct the prehistoric plant environment using clues so small that they are invisible to the naked eye. And his work is so experimental that we don't have any results yet, but it promises to help us obtain facts about prehistoric people which heretofore have been inaccessible.

Back in 3900 B.C., when people built houses in the

Horizon 6 village, we know they used wooden posts for the framework of these structures, because when the posts deteriorated, they left telltale organic staining in the earth. We can still see the imprint of those wooden posts in the ground in the form of round, dark stains going down a few feet below the level of dirt which formed the village floor. But what did they use for roofs? From Irv's work with opal phytoliths, we may be able to determine if they made thatched roofs with grasses.

Opal (a mineral) phytoliths (from the Greek *phyton*, "plants," and *lithos*, "stone") are microscopic structures which have been mineralized in a living plant by progressive silification of cells. Silica and other high-grade minerals are absorbed through a plant's roots. While most minerals are used in the plant's respiratory cycle, silica seems to have no specific function in a plant. It is carried passively through a plant's vascular (water transportation) system and precipitated in or around the cells of root, stem, bark, and leaves. Most frequently, silica is deposited in the wall of a plant's cells. Eventually, because the deposit of silica is continuous, the entire cell may be encased in a cast of silica. When the plant dies and decays, these mineralized casts of cells, which are a bit heavier than pollen and do not float in the air, drop to the soil and remain there where the plant stood. Since the shapes of cells in one plant species differ from those in another species, these casts can be used to identify plant communities as they existed in the past.

Opal phytoliths may prove to be very important in our research, since they preserve better than any other organic fossil and under a wide range of environmental conditions. They also come in large quantities, which is helpful since we need substantial numbers of specimens to reconstruct a prehistoric plant community. Sometimes not enough seeds and pollen have been preserved at a site for adequate analysis.

In addition, in plants such as cereals and fodder grasses, opal phytoliths are more numerous and varied than the pollen. In fact, grasses produce possibly the best identifiable range of opal phytolith types. If Irv can identify grasses in archaeological contexts, this will be an important analytical aid to archaeology for the simple reason that agriculture, worldwide, is based primarily on the domestication of grasses. Corn, wheat, rye, barley, rice, and other grains, as well as sugar cane, are all grasses.

Irv extracts opal phytoliths from soil samples from each horizon of the Koster site by a flotation process similar to that used for plant and animal remains, but employing different chemicals. His work is so new that there are no collections of opal phytolith specimens, no reference texts or sets of reference slides against which his specimens can be compared and identified. His initial task, therefore, is to establish a reference collection of opal phytoliths from known plants in Lowilva, and for this he collects plants from different vegetation zones just as Nancy and Rose do.

Not enough research has been done to make firm conclusions from any of the Koster settlements, but there is one interesting phenomenon. In the soil from sterile levels between the village remains at Koster there is no concentration of opal phytoliths. But in the debris from the different villages and hunting camps opal phytoliths are incredibly abundant, in the tens of thousands per gram of soil. One explanation for this might be that Koster people were using grasses, either wild or cultivated. They may have used them for food, for making thatch for their house roofs, or for baskets or matting.

Eventually Irv may be able to offer proof that the Koster people used grasses in one or more of these ways.

Sometimes, when I watch excavators tossing dirt out of the "big hole," I wonder what other clues to ancient people's behavior we may be tossing out, unknown to us, as opal phytoliths were until recently. I am convinced that as more scientists in other disciplines become aware of what we are trying to do in the new archaeology, they will suggest additional applications of very specialized techniques to increase the amount of knowledge we can derive about the past.

10

Reading the Past Environment From Mussel Shells

One hot August afternoon on my way to visit Gregory Perino, who was digging a burial site on a nearby bluff top, I walked by Macoupin Creek. I could hear voices, and as I approached the creek, I almost stumbled over a pile of mud-caked footwear—hiking boots, Adidas, and assorted other shoes tossed in careless pairs on the ground. In the creek the barefoot owners of the shoes were performing an odd dance, moving each leg in a slow-motion shake while setting the foot down in water.

Sarah Anderson, standing ankle-deep in the water, spotted me and smiled at the quizzical look on my face. "Hi, Stuart. We're looking for mussels. Don't ask me how, but your feet get to know the difference between a mussel and a rock. They're for our reference collection," she said.

The establishment of a reference collection is one of Sarah's major concerns. She works constantly at building up the NAP assemblage of modern animal, fish, bird, and reptile skeletons. To date she has amassed one of the largest faunal comparative collections in North America.

When we are digging, we retrieve faunal remains, such as animal bones or mussel shells, when the earth is tossed through screens. This material is first processed at our central data-processing laboratory in Kampsville, where it is washed and weighed in bulk, then separated into gross categories—fish, bird, and mammal remains and mussel shells. These are placed in boxes and sent to the zoology laboratory. We retrieve smaller items, such as tiny bones and fish scales, in the flotation process.

Very often all the zoologist has is a small fragment of a prehistoric bone. Many bones have what zoologists call "markers" on them, or indicators where there were once muscle attachments; they also have characteristic shapes at the ends. These indicators can be very helpful in identifying ancient bone fragments. By comparing the ancient fragments to modern animal skeletons, the zoologists can vastly increase the number of bones and bone fragments they can identify.

Our zoologists would like to have a number of specimens of every kind of wild animal that existed in North America for the reference collection. The animals in the region today are the same as those that existed during the Koster occupations except for some species that are extinct, such as the passenger pigeon and the Carolina parakeet, and some species that have been forced out of the area by modern people. We try to obtain some of the animals that exist elsewhere by contacting people at other universities or game biologists and working out trades with them.

Sarah collects specimens during different seasons of the year, since certain animal skeletons show different characteristics in different seasons; and for the same reason she collects specimens at different stages of growth.

She carries stacks of large plastic bags in her car in which to place any dead animals she finds along the road. The animals get killed on the highway, and frequently the bones are not damaged. She cleans them off and puts them in an old Maytag freezer on the front porch of the laboratory until she is ready to macerate them. ("Macerate" is the zoologist's word for removing the flesh from the bones.)

News travels fast by word of mouth in Lowilva, and ever since Sarah first began to pick up road kills, local hunters, trappers, and farmers have brought in carcasses as gifts after they removed the skins. Urban ("Cork") Sibley, deputy sheriff of Calhoun County, and his men keep an eye out for road kills for us. They have brought in raccoons, coyotes, squirrels, and muskrats. Once a conservation agent brought us a sixteen-point deer that had been hit by a car.

Sometimes toward the end of summer, after a dry, hot spell, when the creek waters tend to dry up, Sarah will take her students seining for fish for the collection. I watched on one of these occasions as the students walked through the water, dragging the seine close to the bottom. The seine consists of a net of one-quarter-inch mesh about four feet wide and twenty-five feet long attached to a five-foot pole at either end. Under Sarah's direction the crew lifted the ends of the seine and brought them close together. In the net about fifty small fish lay wriggling—sunfish, bass, catfish, and gizzard shad.

"The people who lived at Koster might have caught fish in the very same way we're doing it," Sarah said as the group scooped the fish out of the seine and placed them in buckets. "They might have made nets of reeds or grasses, or they might have made baskets and scooped up fish in those."

Modern animal specimens are prepared by being placed in water and allowed to rot. We use an old abandoned farm for this purpose so that the foul odors won't offend anyone.

Once a week Sarah plans to spend a morning out at the farm. She and her crew remove the specimens from the freezer the night before and thaw them. They weigh and measure each specimen, note the age and sex, where the animal was captured or picked up, by whom, and the date. Each specimen is assigned a number, and when the defleshed skeleton is returned to the laboratory, that number is written on each bone. Out at the farm, they skin the animals and remove some of the flesh, then place them in jars or cans of water and let them macerate.

The modern skeletons are laid out in long, narrow drawers in specially designed cabinets in the zoology laboratory. Fortunately for NAP, the ecology and classification of the animals of Lowilva are very well documented, having been studied by zoologists for many decades, and literature on the fauna of the region is available in libraries through the state. (This is not the case in many areas of the world.)

Each of the NAP zoologists is dealing with faunal material from different sites. Sarah is analyzing evidence from the Crane site, a Hopewell village excavated by her husband, Ken Farnsworth. Dr. Bonnie Whatley Styles, curator of anthropology at the Illinois State Museum in

Springfield, who is both an archaeologist and a zoologist, is analyzing faunal material from Newbridge and Carlin, Late Woodland sites. She excavated Newbridge. Dr. Frederick C. Hill, assistant professor of zoology at Bloomsburg State College, Bloomsburg, Pennsylvania, has been working on the faunal remains from Koster, and it is his work with which I shall be concerned here.

To date, Fred and his crew have analyzed more than 150 species of animals found at Koster. This includes more than fifteen thousand animal remains from the thick, dark-stained soils of Horizons 8 and 6 and the thinner layer of debris from Horizon 4. The job of identifying, labeling, and recording the characteristics of these thousands of bits of material has consumed several thousand working hours over the past seven years.

After they have identified the bones and shells retrieved from Koster, the zoology crew separates the specimens into categories, determined by the environmental habitat preferred by each species. For this purpose, they divide Lowilva into a series of what biologists call biotopes, and which we refer to as resource zones. These are environmental zones characterized by specific populations of animals, birds, and fish.

The presence of various animal or bird or fish species can tell Fred indirectly certain things about the environment inhabited by Koster people. However, the animal remains on a site reflect the biases of the humans who brought them there for food or other purposes. If a species known to have existed during a certain Koster occupation does not appear among the debris from the site, Fred must try to determine whether its absence was due to human choice or to environmental conditions. Some of us, for example, eat pork but do not eat horsemeat, a cultural bias which may puzzle future archaeologists when they dig up our food debris.

Fred's analysis of animal remains from Koster is still in progress. However, his inquiry into changes in the aquatic resources in Lowilva during two Koster occupations is a fine example of how we use new kinds of data to figure out the interactions between human beings and their environment in the past.

When Fred examined the fish and mussel-shell remains from Koster, he found a striking change in the specimens from Horizon 8 times (5000 B.C.) to Horizon 6 times

(3900 B.C.). The majority of fish bones and mussel shells in the Horizon 8 debris were from species that prefer habitats in swift-flowing water such as the Illinois River. By Horizon 6 times, the remains were overwhelmingly from species that prefer sluggish water.

Along with us, Fred found changes in the quantities of waterfowl these two groups of people were consuming. There was little evidence that the earlier people were eating many waterfowl. In contrast, Horizon 6 people were eating lots of ducks, geese, and swans.

Why did Horizon 8 people at Koster concentrate their fishing primarily in the Illinois River, and why did Horizon 6 people, a thousand years later, do much of their fishing in lakes? Was this shift the result of mere human whim, or had it been influenced by a change in environment?

The last glaciers in North America began to retreat about 12,000 B.C., and gradually the climate turned warmer. Between 6000 and 2500 B.C., west central Illinois was undergoing what climatologists call the Hypsithermal Interval, when the area became quite warm and dry. Toward the latter part of this period the climate became more moist and temperate, and in 2000 B.C., it became much the same as it is today. When the region was relatively dry, the Illinois River was "braided," meaning it had several channels which flowed around islands. There were no backwater lakes.

A backwater lake is formed when a river floods. As the river recedes, water is caught in slight depressions in the floodplains and lakes are formed. Some backwater lakes in Lowilva covered several hundred acres. Most were filled with fish, trapped there when the water receded.

Early European travelers in Illinois observed a large number of backwater lakes. A few still existed as late as 1915 (as shown on road maps), although farmers had begun to drain the lakes when they cleared land for agriculture beginning in about 1830. The U. S. Army Corps of Engineers further eliminated the lakes when it began to build levees along the Illinois River to prevent flooding around 1903.

These backwater lakes played an important part in the life cycles of fish and, consequently, in the diet of Lowilva aborigines. Most of the lakes tended to be shallow, and this increased the muddy, reedy areas which support the fauna on which fish feed. The backwater lakes could support larger populations of fish than either the river or secondary streams.

Fred set out to find the reasons for the changes in fishing habits between the Horizon 8 people and their descendants in the Horizon 6 village. As a zoologist, he knew that the composition of mussel shells is controlled by their environment, so he did trace-element analyses of mussel shells from these two time periods.

Specifically, Fred looks at the ratio of strontium to calcium in the mussel shell, because this is determined by conditions in the river or lake in which the animal lives. To study this ratio, the zoologist grinds the shell and dissolves it, then analyzes the residue with an atomic absorption spectrophotometer.

Since strontium and calcium are quite similar chemically and are both absorbed into mussel shells by the same mechanism, they are picked up in whatever proportion is available in the water. The strontium in water is much less concentrated than the calcium. but the exact ratio between the two varies with the conditions. When there is heavy rainfall, the current of the stream quickens, and the water has less opportunity to pick up strontium from bedrock. In contrast, when the stream is slower, the water absorbs more strontium. The animal then picks up from its environment the amount that is there. So the ratio of strontium to calcium will tell you specifically whether the mussel lived in the stream during a period of heavy rainfall or light rainfall. And by analyzing shells from different horizons at Koster, the zoologist can determine whether there was a great or small amount of rainfall during a particular occupation.

The strontium content of the mussel species *Amblema plicata* was much higher in specimens found in Horizon 8 than in similar specimens found in Horizon 6. Therefore we can tell that during Horizon 8 times the river was slower than during the occupation of Horizon 6. Evidently climatic changes brought more rainfall during Horizon 6 times. In addition to diluting the amount of strontium in the river, the heavier rainfall would also have caused the Illinois River to flood, and receding floodwaters would have left behind many backwater lakes. This would account for the larger numbers of fish and mussel species which preferred somewhat sluggish habitats among the Horizon 6 faunal remains.

Horizon 8 people had fished primarily in the Illinois River because that was all that was available to them. A thousand years later, the Horizon 6 people were fortunate to find many backwater lakes, with their large populations of easy-to-catch fish, covering the floodplains. And because migrating waterfowl like to stop and feed on backwater lakes, there were more of these available for the Horizon 6 people.

Thus, by examining mussel shells, Fred was able to learn about climatic changes that took place in Lowilva more than six thousand years ago and had important implications for the Koster people's food supply.

"Take a look."

Dr. Manfred Jaehnig, associate professor of anthropology at Central Washington State College, Ellensburg, Washington, and director of the malacology (shell) laboratory in Kampsville, leaned back and gestured for me to take his place at the double-barreled gray microscope.

There, on a glass slide, sat a tiny snail shell.

"Punctum minutissimum," said Fred. "That means tiny point, and it is absolutely correct, isn't it? This little creature is only 1.2 millimeters across, approximately one twenty-fifth of an inch. It takes four of these little fellows to cover the date on a penny. Not only is he one of the smallest snails, but he is a very slow animal. He has a very restricted experience of the environment; he lives all his life in a ten-foot area.

"But what is even more interesting about this specimen is that he was here a long time before the earliest people lived at Koster. He lived in Illinois forty or fifty thousand years ago, during the Pleistocene (the last Ice Age). We found him in a sample of earth taken from west of Teed's hogpen. There is still Pleistocene loess sand out there. Human beings were evolving in Africa at the time this little fellow lived here."

Gastropods, as snails are known to the scientist, are among the most prolific and diverse animals in the world. Counting both contemporary and extinct varieties, approximately 100,000 species have been identified on earth. These species fall into three major groups: marine snails, which live in the oceans; freshwater snails, which live in streams, ponds, and lakes; and terrestrial snails, which live on land.

At Koster, Fred is studying terrestrial snails. The shell of a snail is preserved if it becomes embedded in non-acidic soil, such as the loess at Koster.

Like people, snails prefer certain places to live on the landscape. Some prefer moist areas, some leaf mold, still others like to stay in warm, dry, grassy areas. By matching the preferred environment of contemporary snails to the statistical distribution of snail shells taken from Koster's occupational levels, Fred can reconstruct what the vegetation and local environment were like at the site in the past.

Diggers at Koster leave a narrow column of earth standing in one corner of each square; a digger bags samples of earth from every three-inch level in the column for snail specimens. At the malacology laboratory in Kampsville, Fred's crew dumps the samples on tarpaper and places them on the grass to dry in the sun. When the samples are dry, they are washed through a fine-mesh screen, and silt, sand, and fine soil particles are washed away, leaving a handful of material.

This residue is dried once again, and then sorted under an illuminated magnifying glass. Students poke gently at the debris with artists' fine-bristled paint brushes, because snail shells are very fragile.

To identify the specimens, Fred examines them under a microscope. Like our other scientists, he too is building a reference collection of modern specimens. In addition, he refers to illustrated reference keys of snails.

One night over a beer Fred confided to me that, like Gregory Perino, he had become fascinated with American Indians after reading James Fenimore Cooper's Leather-stocking Tales, when he was nine, in his native Hanover, Germany.

"It had been translated into German, and as I later found out when I read the English version, they cut down on the descriptive parts, so there was more action in the version I read," Fred said, revealing the remains of a German accent in his soft pronunciation of the *s* sound in "version."

"I wanted to be an Indian after reading that. For Christmas, my dad made me an Indian outfit out of gun-nysack and a feather headdress with turkey feathers which went all the way down the back. He strung pig teeth for a necklace, and he made me a big wooden spear. I had a great time pretending I was an Indian.

"We came to America when I was sixteen, and I hoped to marry an American Indian. But that didn't work out either," he said, laughing. "I married a German-Norwegian American." (Red-haired Jan Jaehnig is her husband's chief assistant in the malacology laboratory.)

And now Fred expresses his affection for Amerindians by trying to reconstruct their prehistoric landscape accurately. For instance he thinks that, long before human beings lived at the Koster site, a lake or pond covered the area for thousands of years. He bases this conjecture on findings from two test squares that reached what appeared to be sterile soils underlying the site, upslope from the deepest part of the site. This soil was a very fine clay which contained large numbers of snails in several very restricted areas. In addition to the snails, Fred found a very large number of *Sphaeriidae*, or pill clams, which exist only in an aquatic habitat. The pill clams, together with the very fine clay material, led Fred to believe that this deposit was laid down by standing water, a lake or pond, which occupied the area long before human beings.

During the Early Archaic period at Koster, when Horizon 11 was occupied, the snails that lived there were species that required no vegetation cover. Fred estimates that the site was open and grassy with a few small trees.

In succeeding levels, from Horizon 10 through most of Horizon 6, these species were replaced by snails that require log and leaf-mold cover, which are characteristic of forest environments or forest openings.

Then, in the uppermost level of Horizon 6, Fred found that snails which require no vegetation cover returned to dominance. From this he concludes that during the periods when Horizon 10 through Horizon 6 were occupied (6000–2800 B.C.), dry and open conditions that had prevailed gave way to increased vegetation that provided more moist conditions for snails.

All of these clues to the Koster Indians' environment show how stable the climate and vegetation were for several thousand years. As I stand on the bluff crests and look out at the Illinois River Valley, I like to think how people have been enjoying this view for at least ten thousand years, and probably longer.

The valley in which the Koster people lived is still very beautiful. Lowilva is largely rural. Farms cover most of the valley floor. Small towns are sparse; some consist of no more than a handful of houses clustered at a crossroads. On a summer day the river floodplains present a palette of different hues of green intermingled with light yellows, the shades reflecting whether corn or soybeans are growing there. Like thick brush strokes on a watercolor painting, there appear the dark grays of macadamized roads, the lighter tans of dirt roads, and the deep brown of the silt-laden river.

Until relatively recently the view of the valley had remained unchanged for more than four thousand years.

Not long ago I stood at the top of the hill behind Teed's white farmhouse with a native of Greene County, looking out over the valley. Kent Feeley, now in his seventies, spent his boyhood on the Koster farm, and he drops by to see us from time to time. He waved a hand out over the floodplains and said, "Those used to be covered with forests when I was a boy."

In about 1820 the first pioneers appeared in Illinois territory to stake out claims to the land. Among them was Robert Clendenin, who built a two-story house of native limestone on his land. The house still stands, just in back of Teed's home and a few hundred yards east of the Koster site. The Clendenins left another memorial on the bluff tops in the form of a tiny family cemetery; two or three headstones still stand, one commemorating a young soldier who died in the Civil War. The Amerindians, too, had buried their dead on the bluff crests, in mounded cemeteries.

Looking out from the tops of the bluffs where we stood, the Clendenins, the Mississippian and Jersey Bluff people of Koster Horizon 1, the Black Sand people of Horizon 2, the Riverton people of Horizon 3, and the Titterington hunters of Horizon 4 all would have seen the identical view.

Where we now see tidy farmlands and neat rows of crops, the Clendenins and the Koster Amerindians looked out over thick floodplain forests. The wet, marshy flatlands near the river were dotted with hundreds of backwater lakes, some of them covering as much as three hundred acres. The Illinois River ran in its present channel, but it was clear and swift-moving then, and its waters must have been a bright blue-green.

The backwater lakes studded the flatlands because the Illinois River flooded a great deal from about 2000 B.C. on. This was partially because the climate at that time had become more moist, but the river's flow also was affected by changes in its headlands, hundreds of miles north, and in changes in the headlands of the Mississippi River, still farther north.

To the earlier villager from Horizon 8 who paused at this spot in about 5000 B.C. on his way to hunt deer in the uplands, the view would have been somewhat different. Then the river ran in several smaller channels, which wound their way about small islands. There were no backwater lakes. Still earlier, in about 6500 B.C., members of the extended family who lived at Horizon 11 would have looked out on a river channel that ran about a mile farther west.

When the earliest people arrived at Koster, about 7500 B.C., the climate was still gradually warming up after the glaciers had retreated. In about 5600 B.C. the climate became much warmer and drier. These conditions persisted until about 2500 B.C., when the climate changed to become moister and cooler; in about 2000 B.C. it became similar to what it is today.

Over the centuries the vegetation in the valley changed only slightly in response to these climatic shifts, because the bluffs served as a buffer to shield vegetation in the enclosed valley against any dramatic changes in climate.

The dramatic changes came in the 1820s, as human beings began to alter Lowilva's vegetation patterns from what they had been for thousands of years. The Clendenins and other settlers began to cut down trees, turn the soils, and drain the bottomlands, to plant crops, and the view which they had shared with their prehistoric predecessors was changed forever.

Thus, when we look at the long record of environmental events in the valley during the Koster settlements, the climate and vegetation altered only slightly over thousands of years. The most radical changes took place in the river and in the land immediately adjacent to it.

One of the results of our studies of the Koster environment has been to open a new page in North American prehistory. Heretofore, archaeologists studying the Archaic people in eastern North America have focused on human beings who lived largely in marginal areas as they interacted with a spreading forest environment after the retreat of the glaciers. Now, for the first time, we are examining in great detail prehistoric groups who lived in a riverine environment, one that was profusely laden with wild-animal and plant food resources.

How people fared in this resource-rich valley, and how they responded to the important changes in their aquatic environment, we'll examine next.

11

Early Organization Man

When we go food-shopping, we take along a list of staples needed to replenish the stocks on our pantry shelves. These staples vary little—salt, sugar, flour, coffee, tea, shortening. We take it for granted the items will be available at the store, although some of them come from thousands of miles away. They are there without our having to think about it, through the intricate workings of a highly organized technological society.

The people who lived at Koster went "shopping" for basic food too; only their list included hickory nuts, cereal-like seeds, white-tailed deer, and lots of fish. Of course, they didn't shop. They hunted and fished and gathered their basic foods, and their "supermarket" consisted of the hillside slopes, forests, upland prairies, floodplains, rivers, and lakes among which they lived.

To our surprise, the Early Archaic people had learned how to exploit the wild-food resources in their environment so skillfully that they could go out and replenish their basic staples on a seasonal basis year in and year out with almost as much confidence as we drive to the supermarket for ours. As a result, they were able to live in great comfort and stability for thousands of years.

It is a revelation to us as archaeologists that there were such highly organized, stable, hunting-gathering societies as early as 6400 B.C., when a small group of people, maybe an extended family, lived in a tiny hamlet at Horizon 11. Prior to Koster, most archaeologists believed that it was not until about 2500 B.C. that Archaic people in eastern North America were this proficient at exploiting wild animal and plant foods.

This evidence also shatters the stereotype that many Americans have about their prehistoric predecessors, that they were brutish people of limited intelligence who barely managed to scrounge enough food from the wilderness to survive.

Our findings from Koster are based on the work of Nancy and David Asch with plant remains and that of Fred Hill with animal remains. When Nancy and David set out to study what plants the Koster people had gathered and used, they not only wanted to learn what the ancient people had eaten but also were curious about how successful they had been at procuring a living and how they had organized their social groups.

To work with, the Asches have only those parts of a plant that have been tossed or have fallen into a fire and become carbonized. The leaves and stem of most plants burn completely, but there are some inedible by-products of plants that people discard, such as nutshells, stems of apples or pears, pits of certain fruits, and seeds, which, when exposed to fire, will survive as carbonized remains. Nutshells, being hard and dense, burn slowly, and the charred fragments of these can withstand rough treatment. Charred bits of seeds of some plants also can be recovered at archaeological sites.

When Nancy and David examined the plant remains from the little hunting camp at Horizon 4 and the villages at Horizons 6, 8, and 11, they found the bulk were charred nutshells. Charred seeds were preserved in much smaller quantities, indicating a greater dependence on nuts during those occupations.

We know that early people ate seeds because remains of seeds (including those of marsh elder, *Iva annua*, and goosefoot, *Chenopodium*) have been found at sites in Kentucky in human feces (referred to as coprolites in archaeological studies). No coprolites have been found at Koster, but we have found charred seed remains there and at other sites in Lowilva, and we can assume that people were using these as food.

As scientists we can deal only with that which has survived as evidence, which is why, in this discussion, you'll find us talking only about charred shells and seeds. Koster people lived in one of the most luxuriant areas of the continent. We can assume that their diet included many edible wild plants, just as that of their descendants did thousands of years later; but without direct evidence, we can't prove it.

Sometimes we use ethnographic analogy; that is, we look to today to understand yesterday. We study a living culture, or one which was observed during historic times. Then we compare the known culture with a prehistoric one which lived in a similar environment or shared similar characteristics.

When Europeans encountered the Amerindians in the Great Lakes area (some of whom would have been the descendants of Koster residents), they were astounded at the large number of plants the aborigines used. The Amerindians were observed using 275 species of plants for medicine, 130 for food, 31 as magical charms, 27 for smoking, 25 as dyes, 18 in beverages and for flavoring, and 52 others for various purposes.

Today we tend to ignore the possibilities of wild plants as foods because we think only in terms of cultivated plants. Many of the plants we regard as useless or annoying weeds are edible, tasty, and nutritious. Historically, Amerindians in Lowilva were observed eating tubers of duck potatoes (*Sagittaria platyphylla*), which taste surprisingly like our own cultivated potatoes, and water lotus (*Nelumbonucifera*). Goosefoot is a weed that grows profusely in the Middle West; it grows in wild stands next to the Koster site. The late wildfoods expert Euell Gibbons wrote that his sister picked goosefoot, parboiled and froze it, just as commercial food processors freeze cultivated greens. Nancy cooked some goosefoot for us; we found it very tasty, somewhat like spinach. Probably the Koster people tossed goosefoot into their stewpot, along with other ingredients.

Our students gain insight into the prehistoric people we study by replicating the life-ways of those ancient cultures. Each fall Nancy and David, together with Rose Duffield, take students enrolled in the NAP field school on a wild-foods hunt. They spend several days fishing and collecting wild plants, and occasionally a local resident will contribute wild game. The students prepare the wild foods, and everyone in NAP gets together to eat these at a picnic. We invite some of our Kampsville neighbors to join us. Students and guests who are trying these foods for the first time always are impressed at how delicious a meal made exclusively from wild foods can be. Many students, having had a glimpse of what a good life might have been led by the Koster Amerindians despite the latter's simple culture and technology, come away from this experience with a heightened respect for their prehistoric predecessors.

During most of the major occupations (when there were very large villages at Horizons 8 and 6 circa 5000–2800 B.C., and even when just hunters were camping at Koster during Horizon 4 times, 2000 B.C.) hickory nuts were the major food for which evidence can be found. Koster people also ate black walnuts, acorns, pecans, and hazelnuts, but apparently in much smaller quantities than hickory nuts.

Mixed in with the food remains, the Asches found lots of charcoal bits of walnut and oak, indicating that these were being used as firewood. Apparently there would have been plenty of walnuts and acorns available from these trees, but the Koster people chose to collect hickory nuts as their prime food source instead.

This choice would have earned the Koster women (who were the cooks in their society) high marks from nutritionist Adelle Davis, who preached the need for well-balanced diets for good health. Hickory nuts are easily digestible and have a high protein content which resembles the meat of large mammals. They are a good source for high energy. Since the Koster people ate a lot of freshwater fish, which is very lean, their choice of hickory nuts was a very wise one; the nuts provided an excellent dietary complement to the lean fish.

Nancy, as a scientist, sometimes goes to extraordinary lengths to find answers to questions about the Koster people which puzzle her. One day, over tea in her laboratory, she said, "Stuart, do you know that when we went through the plant remains from the village at Horizon 11, and a few of the squares they've opened up in Horizon 12, we found those people much preferred pecans in their diet? They ate a few hickory nuts, but not nearly as many as pecans. Yet from about 5000 B.C. on, in all of the other occupations, the overwhelming choice was hickory nuts over all others. Why do you suppose there was such a difference?"

As we talked, David joined us, and we tried to figure

out this difference in exploitation of an important food staple.

"Maybe in Horizon 11 times people lived at Koster only during certain seasons of the year and were living elsewhere when the hickory nuts were ripe," suggested David. "The pecan crop in Lowilva is concentrated in November, after most other nuts have fallen and been consumed by squirrels and deer. They may have been off somewhere else on hunting or food-gathering trips, using Koster as a home base, and may have arrived back during the time when the pecan crop was ready."

"Yes, but that leads to other questions," I said. "If the people during Horizon 11 times did not live at Koster in the late summer and early fall, they must have been following an annual settlement pattern different from that of the people who lived there later, during the times when there were very large villages. From the evidence, it looks as if the big villages at Horizons 8 and 6 were occupied all year round."

"I've been considering other possibilities," said Nancy. "Maybe there were changes in nut crop production in Lowilva over a period of time due to changes in the climate. Ken [Farnsworth, an NAP senior archaeologist] thinks the Illinois River channel might have been about a mile closer to the village during Horizon 11 and 12 times than it was a few thousand years later in Horizon 6 and 8 times. If it was closer, then the floodplain forests would have been closer, too, and it would have been easier to collect pecans then than to walk farther to collect hickory nuts.

"I'm planning to do a study of the nut yields in different places in Lowilva to test some of these hypotheses," said Nancy. "Do you think you could lend me the pickup truck tomorrow? I can't fit the entire class [in the field school] in our car."

Nancy planned to drive to Red's Landing, an Illinois state wildlife preserve along the Mississippi River about nine miles west and south of Koster. When the next day dawned soft and golden, I decided to go along.

As we pulled into Red's Landing, Nancy waved at a grove of trees and said, "We're going to do a count on those shellbark hickory trees."

Within a few moments Nancy and her students had scattered, each one taking a station beneath a different tree. Some sat on the ground, some stood. A few held binoculars to their eyes. All peered intently into the canopies of the trees. Occasionally a watcher would pause and jot down some notes on a pad.

To make the count, each person chose four branches at random on each tree, two branches at the top and two at the bottom, and counted the nuts on those. Nancy's object in doing the nut count is to figure out what the typical yield of nuts per tree is for the season. She wants to estimate the differences in nut yields between different areas in the region.

Eventually Nancy hopes to be able to determine the dependability of various nut resources—among them, hickory nuts, walnuts, acorns, and pecans—in Lowilva over a span of years. She will then translate her findings into estimates of how much of these foods might have been available on an annual basis to Koster residents.

For example, shagbark hickories produce nuts only every other year; white oaks produce acorns only once in every seven to eleven years. Do the hickory trees in one stand produce on a staggered basis, with some producing one year and some another year, or do they all produce nuts one year and none the next year? And are nut crops of single species synchronized over a whole region, or do single trees or groves of trees produce nut crop yields at random?

When their counting was done, Nancy and her students placed a small metal label at the base of each tree; she planned to come back to these same trees over a number of years and count the nut yields.

She has chosen this spot and others at several Illinois colleges because she thinks there is a chance that these hickory trees will remain undisturbed, whereas trees on much of the property in the area are being timbered by farmers for profit.

Koster women must have worked very rapidly to harvest large numbers of nuts each year to beat the squirrels to the crop. (In most nonagricultural societies, women traditionally are the food-gatherers; men are the hunters.)

Among the charred seed remains taken from the narrow band of stained earth which constituted Horizon 4, and from the thick, charcoal-colored, debris-laden soils of Horizons 6 and 8, Nancy and David found that the majority were marsh elder.

Artist's rendering of a Helton village in the Lower Illinois
River Valley, circa 3500 B.C. (JAY MATTERNES)

Above: An aerial view of the Koster site, showing the main trench in 1975. The floor of the trench as seen here is in Horizon 11, the remains of a village dating to 6500 B.C. (D. R. BASTON)

Right: Gail Houart, Koster site supervisor, and Tom Styles remove a soil sample from below the deepest Koster horizons to determine whether there are more occupation layers buried underneath present excavation. Excavators use a soil auger drilling rig, mounted on back of a truck, to retrieve samples this way. (D. R. BASTON)

Left: Harlin (Alec) Helton holds a prehistoric pot, one of his collection accumulated over more than fifty years of prehistoric-artifact-collecting in the Illinois valley. Helton is one of the discoverers of the Koster site, bringing it to Stuart Struever's attention in 1962. (GARY REYNOLDS)

Below: The main excavation at the Koster site in 1976. This photo shows the immense size of the main excavation at Koster, one of the largest and most complex excavations ever undertaken in North American archaeology. (D. R. BASTON)

Top: One of the surprises of the Koster site was the discovery of skeletons of domesticated dogs. This is one of several dog skeletons discovered in a "canine cemetery" in Horizon 11, the remains of a village carbon-14-dated at 6500 B.C. (FRANKLIN McMAHON, JR.)

Bottom: Theodore (Teed) and Mary Koster, with their German shepherd, Gypsy, at the famous site that bears their name. (D. R. BASTON)

Left: Stuart Struever stands on the floor of the main excavation at the Koster site in 1977. (BILL HAGEN)

Below: Samples of pollen, showing how distinctive each can be. These enable palynologists to identify which plants grew at the site thousands of years ago. (SMITHSONIAN INSTITUTION)

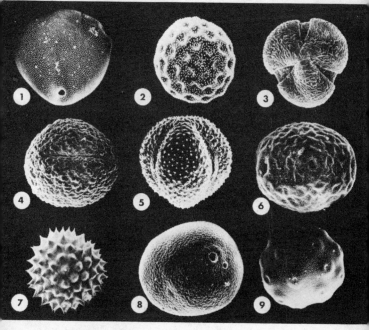

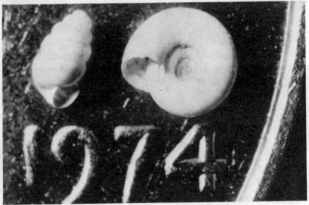

Top: A student identifies animal bones
found at Koster in the zoology laboratory in Kampsville.
(D. R. BASTON)

Bottom: Two snails are magnified on a
penny. This photo reflects the tiny snails that are often
recovered and identified at Koster. (D. R. BASTON)

Top: Bits of burnt clay representing "daub."
Note the woven material or stick impressions in the
surfaces of these burnt clay fragments. Are these
evidence of early plaster-wall houses dating to 3000 B.C.
at the Koster site? This is one of the problems that
the Koster archaeologists hope to solve in the
years ahead. (FRANKLIN McMAHON, JR.)

Bottom: A close-up of material retrieved from
the living floor at Koster after it has been washed
and separated. Lower left shows clam, mussel, and
snail shells; lower right shows pieces of broken chert
artifacts and bits of chert; upper right shows
pieces of limestone. (D. R. BASTON)

Above: Some of the many styles of projectile points —used as spear tips—found at the Koster site. (D. R. BASTON)

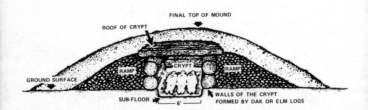

FINAL TOP OF MOUND

ROOF OF CRYPT

CRYPT

RAMP

RAMP

GROUND SURFACE

SUB-FLOOR

WALLS OF THE CRYPT FORMED BY OAK OR ELM LOGS

6'

Opposite: This diagram shows a cross section
of a typical Hopewell burial mound in Lowilva. The
highest-ranking people were buried in the place of honor
in the central log crypt. The majority of these
were males, although occasionally a female was buried
with males in the crypt. Lesser personages were
buried in the earthen mound surrounding the crypt.
(GEORGE ARMSTRONG)

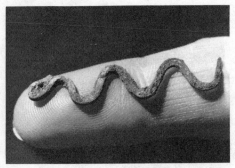

The serpent was an important
religious symbol throughout prehistoric
North American cultures. This one, made
of copper, dates back to the Mississippian
period (A.D. 1000-1673). The serpent
may have been part of a shaman's tool
kit. This one was excavated at the Carter
Mound Group, a burial site about ten
miles from Koster. (D. R. BASTON)

Dr. Jane Buikstra, archaeologist and
biological anthropologist, uses a paint brush
and Perino pick to clean away dirt from
the skeleton of a person buried two thousand
years ago in the Helton burial mounds.
(ROBERT PICKERING)

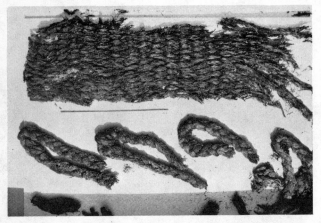

Above: A close-up view of a piece of 1,400-year-
old fabric woven by the White Hall people (circa A.D.
400-600) from plant fibers and bark. (D. R. BASTON)

Opposite: A profile of the north wall at the Koster
site. The lighter-colored layers are sterile soil, laid
down by "slope wash" washed down the bluffs by rain
while the site was unoccupied. The dark layers are the
debris-laden soil left when people lived there.
The darkest layer (with what look like chunks of rock
sticking out) shows debris from Horizon 6 (circa
3900-2800 B.C.), the most densely occupied stratum at
the site. The dark layer near the bottom, to which the
student is pointing, is debris from Horizon 8
(circa 5000 B.C.) (D. R. BASTON)

Top: View from the floor of a test square at the Koster site. This photo highlights the great depth of the Koster site, which contains the remains of prehistoric Indian villages dating from at least 7000 B.C. to A.D. 1000. These village remains show up as the dark-stained layers in the wall being shovel-scraped by a student excavator in this photo. (FRANKLIN McMAHON, JR.)

Bottom: Stuart Struever examines the remains of an infant recovered from Horizon 11. The skeleton was so fragile that a plaster casing was placed around a block of soil containing the remains and the entire block, with skeleton in place, was then removed to the laboratory for closer examination. (D. R. BASTON)

A student excavator uncovers the remains of a human skeleton. Note the superb preservation of the bones, though this burial has been in the ground for almost five thousand years. (FRANKLIN McMAHON, JR.)

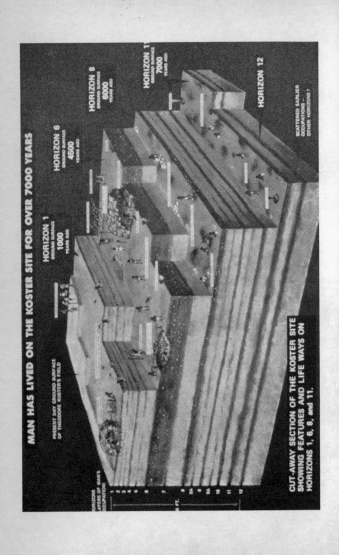

MAN HAS LIVED ON THE KOSTER SITE FOR OVER 7000 YEARS

CUT-AWAY SECTION OF THE KOSTER SITE SHOWING FEATURES AND LIFE WAYS ON HORIZONS 1, 6, 8, and 11.

PRESENT DAY GROUND SURFACE OF THEODORE KOSTER'S FIELD

HORIZON 1
GROUND SURFACE
1000
YEARS AGO

HORIZON 6
GROUND SURFACE
4500
YEARS AGO

HORIZON 8
GROUND SURFACE
6000
YEARS AGO

HORIZON 11
GROUND SURFACE
7000
YEARS AGO

HORIZON 12

SCATTERED EARLIER OCCUPATIONS — OTHER HORIZONS ?

Marsh elder is one of two plants (the sunflower, *Helianthus*, is the other) we believe were being cultivated by Early Woodland times in Lowilva (500 B.C.). When plants are cultivated, they sometimes produce larger seeds than those of wild varieties. Seeds of marsh elder larger than those of any other species or variety of the plant, including those of present-day wild size range, have been found at archaeological sites at the Ozark Bluff Dweller sites in Arkansas and Missouri. Might the people who lived at Koster in Horizon 6 times have been cultivating marsh elder?

Marsh elder grows extensively on wet bottomlands and is not likely to be found growing naturally on a hillside location like Koster. Some goosefoot species, by contrast, are admirably adapted to the drier, disturbed, nitrogen-rich habitat that would have existed on the site when people lived there. (When one walks on the edge of the site today, one is apt to brush against stands of goosefoot.) Goosefoot produces hundreds of seeds per plant; the number of goosefoot seeds accidentally dropped into campfires at Koster (which turn up in the flotation process as charred remains) must have been astronomically higher than those of marsh elder. Yet in the archaeological samples from the large village at Horizon 6 there are more charred seeds from marsh elder than from any other plant.

This suggests that the Helton people at Horizon 6 were harvesting marsh elder seeds in large quantities, just as today we harvest seeds of cultivated wheats or oats. It was an extremely clever adaptation. The Helton people reaped a harvest without having had to work for it other than shaking the stalks of marsh elder plants over the baskets in which they collected the seeds. And this excellent food source was annually self-renewing; no one had to place an order for it, package it, ship it, or stock it on grocery shelves.

We think the Helton people traveled to wet bottomlands to gather these seeds, or they were growing marsh elder near the site.

In the fall of 1972 Nancy and David were conducting some experiments with marsh elder, trying to learn something about the potential productivity of the plant by going out and harvesting and measuring the weight of achenes (outer seed coats) per square meter plot.

They had stored their harvest over the winter in the

botany laboratory. In the spring of 1973 the largest flood that ever occurred historically in Lowilva washed most of their harvest out the window. By June the flood had receded; it was a very dry summer, and soon the soil around the lab became baked out. But the Asches noticed strange little plants growing in the vicinity of the lab. These plants remained a few inches high until the end of July, when suddenly they had a very rapid growth.

They realized they were looking at the growth of marsh elder, in very dry soil, right near the laboratory. How could we explain this, since normally it grows only on wet bottomlands?

Marsh elder germinates later than other plants, and its growth is related to day-lengths (i.e., the amount of light available to it according to the length of daylight each day). The Asches realized that some marsh elder plants had been sitting next to the laboratory during the months of June and July unnoticed, because they remained only a few inches high while other weeds around them had a tremendous growth spurt during that time. When the marsh elder was next to other weeds, it was shaded out and did not thrive.

By chance, the Asches had seeded some marsh elder into a lawn, and it was able to hold its own in terms of light requirements with the short grasses next to it.

This may very well explain what happened in prehistoric times in terms of human manipulation of marsh elder. It is very easy to see how the Amerindians could have drastically increased the productivity of marsh elder simply by controlling the competition with other plant species during the early part of the season.

Several years before I discovered Koster, I had excavated the Macoupin site, a village which was occupied during the Middle Woodland period (circa 100 B.C.–A.D. 450). People lived at Macoupin just about the time Koster Horizon 2 was being abandoned (100 B.C.), and remained there until some time after people had begun to establish a large town at Horizon 1 at Koster (A.D. 400–1000). Macoupin may have been a part of one of the Koster settlement systems. At Macoupin we had recovered hundreds of bits of charred seeds in the food debris.

When Nancy and David compared the food remains taken from the Macoupin site with those from the village at Koster Horizon 6, they found that the ratio of seeds to nuts from Macoupin was thirty-eight times the corresponding ratio of seeds to nuts found at Koster. This told us that the Macoupin Middle Woodland people exploited seeds as an important part of their diet, whereas the Koster Archaic people had used them as a minor food supplement.

When the Asches compared the size of the marsh elder seeds at Macoupin with those from the Koster debris, they found the ones from Macoupin were larger. This factor, together with the increased number of seeds from Macoupin, suggests strongly that by Middle Woodland times the Amerindians were cultivating marsh elder in Lowilva.

Chance had also played a role in our discovery of other evidence for the striking difference in the use of seeds between the Archaic and Woodland peoples in Lowilva. In 1962 I was digging the Apple Creek site. The expedition consisted of myself and a handful of students.

One day, we had toiled in hundred-degree heat and, after showers in our home-made outdoor stalls, had just sat down to a hot meal, when there was a knock at the door. The caller was a highway contractor who had visited the Apple Creek site that day and had seen the storage pits that were scattered all over the site. He explained his mission. In excavating soil for the new blacktop road between Eldred and Hillview, the road crew had encountered a large number of dark circles in the floor of the borrow pit next to the roadbed. The contractor recognized them as being similar to the storage pits at Apple Creek. He suggested that, if we were interested in examining these pits, we do so immediately, since the road crew would be bringing in earth-moving machinery early next morning.

I glanced at my watch. It was six-thirty and darkness would not come until almost nine o'clock on this summer night. I looked longingly at the hot food on my plate, then made my decision, saying to the students, "O.K. These pits will be destroyed in the morning. It sounds to me as if it is a Woodland habitation site with big storage pits in it, and it's just a little bit north of Apple Creek. What do you think we should do? I'm going out there to dig. If any of you want to join me, come along."

The entire crew got up, leaving the hot food untouched. We dug at the site until it was too dark to see.

The night's labor yielded a very important find. There were eleven or twelve pits, which had been used for storage. In two of these we found masses of charred seeds. We had never before found food remains in such quantities to document early people's consumption of seeds and their storage techniques.

A couple of years later, after Nancy had joined NAP, she set about the laborious task of counting and identifying these seeds. She reported that the pits had contained 4,700,000 charred smartweed and goosefoot seeds.

Charcoal from that site (which we called Newbridge) was later radiocarbon-dated, showing it to be what I had guessed it was, a Late Woodland site, circa A.D. 700.

So, from the food remains taken from the Koster villages and from the Macoupin and Newbridge settlements, we can trace the gradual increase in the use of wild seeds as an important staple in the diet of prehistoric people in Lowilva. In 3900 B.C., although Horizon 6 people were harvesting seeds as annual "crops," wild seeds were less important in their total diet than hickory nuts. By 100 B.C. at Macoupin people were eating much larger quantities of seeds, probably cultivating one species, marsh elder, to increase the yields. Then by A.D. 700 Newbridge people were consuming massive quantities of seeds in comparison to the amounts that people ate back in 3900 B.C. By this time the population in Lowilva had increased, and we suspect that as the residents were forced to find additional food sources, they included more seeds as a dietary supplement.

And what kinds of meat and fish did the Koster people find when they went to their "butcher shop," the forests, fields, and waters of Lowilva?

In the seventeenth century Father Rasles, a Jesuit priest who lived with the Illinois Indians, wrote about them: "Among all the Tribes of Canada, there is not one that lives in so great abundance of everything as do the Illinois. Their rivers are covered with swans, bustards, ducks, and teal. We can hardly travel a league without meeting a prodigious multitude of Turkeys, which go in troops, sometimes to the number of two hundred. Bears and deer are found there in great numbers; there are also found countless numbers of oxen [bison] and of roebucks [deer]; there is no year when they do not kill more than a thousand roebucks, and more than two thousand oxen; as far as the eye can reach, are seen from four to five thousand oxen grazing on the prairies."*

This incredible abundance of animals apparently existed all through the prehistoric occupations of Koster, with one notable exception. There were no bison bones in any of the Koster middens. If bison had been plentiful when Koster was occupied, then we would expect to find their bones at the site. We know that Koster people's technology was sufficient to kill bison, since bison bones embedded with stone projectile points have been found elsewhere in North America. So, because there were no bison bones at Koster, we assume they were not in the area. It appears that bison were very recent arrivals in this part of the Middle West, and in fact bison bones have not been found at any archaeological sites in the area which predate A.D. 1500.

We are trying to guess why the bison appeared in Illinois at that time. Does it reflect a shrinking of forests and expansion of the prairie, giving the bison a more extensive habitat? If so, is this the result of human groups burning off areas of the prairies, perhaps for agriculture? If you burn off the prairies around forests, you can cause the forests to retreat, and this may account for the appearance of bison in Illinois at this late date, when agriculture was being practiced by the Amerindians.

Koster people had a diet rich in animal and fish protein, apparently taking with ease all they needed from among the numerous animals, birds, and fish from the surrounding forests, prairies, rivers, and lakes.

Throughout the four-thousand-year period when there were major occupations at Koster (circa 6500–2000 B.C.), the most important animal staple was the white-tailed deer, the third largest animal in the environment, next to the elk and the black bear. But Koster people also ate other mammals, including raccoons, woodchucks, beavers, muskrats, cottontail rabbits, squirrels, and dogs. They enjoyed waterfowl such as ducks, geese, and swans, as well as prairie chickens and wild turkey.

Fish were a very important part of their diet, as were freshwater mussels. Among the fish they ate were bass, buffalo fish, bowfin, catfish, sunfish, and freshwater drumfish. The amount of fish remains came as a surprise. Ap-

* Ruben Gold Thwaites, *The Jesuit Relations and Allied Documents* (Cleveland: Burrows Bros. Co., 1896–1901).

parently Early and Middle Archaic people included a great deal of fish in their diet. It had previously been assumed that during this long period (8000–2500 B.C.) the aborigines had lived primarily on land mammals and a few plants. But the Koster evidence tells a different story. As early as 6500 B.C. they were eating enormous quantities of freshwater fish and mussels.

The most fascinating revelation from the fish remains was that as early as 3900 B.C., when the first of a series of large villages was established at Horizon 6, Koster people were harvesting fish like we harvest plant crops, taking vast numbers at a time. As I mentioned before, these people were doing most of their fishing in backwater lakes. The fish bones left behind by the villagers at Horizon 6 and the hunter-campers at Horizon 4 ranged in size from the tiniest fingerlings to the largest species which thrive in backwater lakes. The Amerindians had been taking total populations of fish in every size and every age range from these lakes.

We also found fish bones that were smoke-blackened. Apparently Horizon 6 people were smoking some of their fish harvest and storing the dried fish for use during the lean winter months, when fresh foods would have been scarce.

These were the same people who were the first to gather wild seeds in large quantities. The collecting of these readily available, annually renewable food sources—wild seeds and freshwater fish—and the processing of them for storage over the winter was the Koster people's ingenious way of stocking their own pantries from nature's bounty.

These new findings can be credited to our flotation process. We are floating out hundreds of small fish bones that used to go unnoticed at archaeological sites. They also reflect the superb preservation of the tiniest animal and fish bones in Koster's nonacidic soil.

We have no evidence for the techniques Koster people used for fishing, but we can speculate about their methods by looking at those of historic hunter-gatherers, or even our own. To obtain so many fish at one time, they may have been collecting fish in nets or baskets. They may have seined for fish by dragging large nets held by several people along the bottom, as we have in an attempt to imitate prehistoric fishermen in the Illinois River and Macoupin Creek. On one occasion, we caught as many as three hundred small fish within a half-hour. "That beats any labor they would have had to put in for planting, weeding, watering, and hoeing crops," remarked Sarah Anderson.

Or they might have poisoned the fish by tossing crushed hickory-nut shells in the water, as historic Amerindians did. The fish die and float to the top.

The aborigines also may have used a technique which President Jimmy Carter used as a boy in Georgia, described in his book *Why Not the Best?* He stirred the mud up from the bottom of a pond or lake with his bare toes to deoxygenate the water; the stunned fish would float to the top, where they could be scooped up.

Dr. Andreas A. Paloumpis, a fisheries biologist and a personal friend, has studied the ecology of fish in the large river valleys of the American Middle West, and he estimated that the Koster people may have been able to take from three hundred to six hundred pounds of fish flesh per acre of water in the backwater lakes. When you consider that some of these lakes covered as much as several hundred acres, you can see that harvesting the "crop" from just one lake would have been a very productive venture for even a small group of people. The fish populations in the Illinois and Mississippi River Valley floodplains may well have made these two areas among the most productive environments for thousands of square miles.

We also use the plant and animal remains to help us determine what seasons of the year people were living at Koster, in order to learn more about how they adapted to their environment. For example, if a site yields a great many hickory-nut shells and very few fish remains, it suggests that people lived there only in the fall when nuts were ripe and it was too cold to fish.

By figuring out which seasons of the year the Amerindians lived in different parts of their environment, we can trace their settlement systems.

Fred Hill examines prehistoric fish scales for clues to seasonality (and for other kinds of information). The prehistoric scales are very brittle and difficult to handle without breakage, so he makes a print of each scale on a plastic microscope slide to study it. Then he places the slide into a microscope enlarger, which projects the image of the scale on a screen, magnified about forty times.

Across the surface of the magnified fish scale one can see concentric rows of arcs of alternating widths. Just as a tree adds rings to its trunk each year, the fish scale adds an arc, called an *annulus,* which usually is laid down in late winter or early spring. When the annuli are laid down, they interrupt the growth of the rest of the scale; the spaces between the annuli are called the *circuli.* The two appear to be of different textures. Seen on the screen, these thick, concentric arcs, annuli and circuli, suggest the whorls on a human fingerprint. They occur only on the outside surface of the scale.

By counting the number of annuli on a fish scale, Fred can tell the age of the fish at capture, which lets him know whether Koster people were being selective about fish they took or whether they were taking all ages (and sizes) in a fish population. By measuring the distance from the last annulus to the scale's edge, he can tell in which season of the year the fish was caught, and from that infer what seasons of the year people were fishing in nearby streams or lakes. And by measuring the amount of growth each year, as revealed by the breadth of the annuli, he can tell whether the growing seasons had been good or bad during the fish's lifetime. "Good" or "bad" means whether or not the fish was receiving the right nutrients from the stream; it gives Fred clues to what was happening in the fish's environment at the time. He also counts the annuli on freshwater mussel shells.

For further clues Fred examines deer teeth and antlers. To determine the age of a deer at capture, he looks at the teeth. In a preadult deer the rate of replacement of baby teeth by permanent teeth occurs at a known rate (from studies of modern deer). In an adult deer Fred examines the degree of wear on the teeth. Modern hunters are required to bring killed deer into a game station in most states; from these specimens studies have been made on deer jaws. Fred refers to these studies in trying to determine the age of an archaeological specimen. He also uses deer jaw specimens from the zoology laboratory's comparative collection as a reference.

White-tailed deer are born at approximately the same time of the year, in May or early June. The deer antler replacement has a regular cycle of three distinct stages—no antlers, a soft stage, and a hard stage—each of which lasts for about four months. Thus, if a deer skull has antlers, the animal was probably killed in fall or winter. If it has no antlers, it was killed in spring or summer. Most of the samples from Koster, Fred found, had been killed in fall or winter.

Fred's and Nancy's analysis of animal and plant remains suggests that people lived at Horizon 8 (5000 B.C.) throughout the major portion of the year, with the possible exception of the winter season. Fred's fish-scale analysis suggested that people were at Koster from late spring through summer; and the analysis of freshwater mussel shells placed them there through the fall. The nutshells also suggest early fall occupation.

The story seen from the animal and plant remains from the large village at Horizon 6 (3000 B.C.) is a different one. Animal remains indicated that this village was occupied all year round. From the deer antlers Fred inferred they had been at Koster between September and January; from the deer teeth between August and January. Migratory waterfowl remains indicate spring and fall occupation, and the large number of freshwater mussels suggests summer or fall occupation. Finally, the fish scales suggest that the area was occupied throughout the entire year. From their food-gathering patterns and from the seasonal indicators we can see that Koster people had managed to become quite settled, living for long periods in one place as far back as 5000 B.C.

The discovery that these people had achieved such a high degree of stability so early has caused us to change our views about certain cultural developments among prehistoric people of eastern North America.

Earlier, archaeologist Joseph Caldwell suggested that it was not until 2500 B.C., in the Late Archaic period, that Amerindians had become expert at exploiting animal and plant resources in eastern North America. He theorized that even in environments with an abundance of wild foods people can have difficulty getting along if they don't know how to obtain and use the foods available. He thought it took people most of the Archaic period, from about 8000 B.C. to 2500 B.C., to develop efficient techniques for collecting wild foods, and he coined the term "primary forest efficiency" to mark the time when at last people achieved this state.

Caldwell based his conclusions on the knowledge then available about the Archaic period. Koster, with its stackup of so many intact settlements over such a long time span, has enabled us to take a new, in-depth look at the period.

By 5000 B.C. Middle Archaic people at Horizon 8 were extremely well adapted to their environment and were obtaining delicious, nutritious foods easily, in amounts which later people were to achieve only through agriculture.

Besides, why would a population depending solely on wild plants and animals have required thousands of years to learn the efficient use of these resources? The presence of acorns, black walnuts, hazelnuts, and pecans at Koster indicates that Archaic people were familiar with these resources but chose not to exploit them in large quantities. And the relatively small quantities of edible seed remains at Koster does not mean the people were ignorant of the food values of this resource; they simply did not need to collect more seeds because they had so much other food available.

Koster Archaic people were adapting very efficiently by concentrating on a narrow range of excellent foods. Nancy refers to what she calls "first-line" foods—those which are abundant, easily collectible, and most nutritionally complete. The Koster people were able to subsist very well by focusing on the collection of just a few first-line foods, among which hickory nuts are a prime example.

It's interesting to note that walnuts are as nutritious as hickory nuts and also grow abundantly in the area, but they are not as easy to harvest. Walnut trees tend to be widely spaced throughout a forest, because their roots contain a hormone, *juglone*, which inhibits growth of other walnut trees. And when you pick a ripe walnut, you must remove the husks by hand. In contrast, the ripe husks of hickory nuts split naturally. Koster people chose hickory nuts over walnuts because they were easier to gather and prepare.

Our data also challenges some of the theories of Thomas Malthus, the eighteenth-century English economist, who believed that human population size quickly grows to consume more food, and therefore is directly limited by the food supply. Since there appears to have been a stable adaptation at Koster for at least three thousand years (5000–2000 B.C.), during which Middle Archaic people concentrated on a narrow range of first-line foods that could be gathered and processed easily, we think the human population changed only slightly in size and didn't need to place serious demands on the plentiful natural foods in their environment.

Dr. Lewis Binford has pointed out that evidence is lacking for two corollaries of Malthusian theories. Malthus thought that people would continually be seeking means for increasing their food supply. The Koster findings show that isn't true.

And Malthus, along with Caldwell, thought that people lack the incentive to become creative until they can stop worrying about where to find more food.

Again, the Koster findings dispute both Malthus and Caldwell. If you look at many present-day hunter-gatherers, such as the !Kung San in the Kalahari desert of Botswana, Africa, you find they have a good deal more leisure time than we do as members of industrial societies, yet they choose to pursue the same life style they have had for centuries.

Koster hunter-gatherers probably enjoyed much leisure time too, since the prime sources of food were all to be found within a three-mile radius of their villages. And the supply was plentiful as long as the population remained stable. Yet their societies appear to have remained relatively simple in structure, and there was no great flowering of culture and creating of beautiful objects until later, in the Middle Woodland period (100 B.C.–A.D. 450). We think it is competition which drives people to become more creative, rather than the availability of leisure time.

During the Late Archaic and Woodland periods some important cultural changes took place in Lowilva. In the Late Archaic period, about 2500 B.C., the population began to grow; there was another increase in Middle Woodland times, (100 B.C.–A.D. 450), and by the Late Woodland period (A.D. 450–1000) the population in the region had expanded again and was considerably larger than it had been during most of the Koster occupations. It continued to grow from then on, through Mississippian times (A.D. 900–1673), up to the time the first French explorers arrived. As the population grew, people were forced to settle in areas where the prime food resources were less plentiful. Eventually people began to experiment with cultivated plants, and by A.D. 900, in the Mississippian period, Lowilvans had become largely dependent on agriculture to procure most of their food. Along with this shift in their economy, they developed larger towns and more complex social structures, and produced artifacts in larger numbers than before. And, as I have mentioned, during the Hopewellian period the artifacts were not only useful but beautiful in design.

The conversion from hunting-gathering to agriculture is one of the most important changes a society makes as it goes from simple to complex. In most societies around the world, each culture, after making that shift, has continued to grow in complexity. Consequently, we are intensely interested in trying to understand the processes which caused simple societies to initiate the steps that would eventually transform their cultures so completely, and for all time.

Many archaeologists have assumed that, in making this shift, people first sought ways to add to their food supplies and then increased the size of their populations.

In Lowilva this important cultural process took place in reverse order. First the population expanded, and then people were forced to find ways to increase the food supply.

British anthropologist Ester Boserup, in conducting studies of agricultural growth in modern underdeveloped countries, concluded that human beings do not necessarily seek higher levels of productivity without a good reason for doing so. She saw the pressure of increasing population as the chief stimulus for a shift from hunting-gathering to agriculture as a subsistence base.

In fact, the abundance of animal and fish resources available to Koster people was such that it may well have delayed the development of agriculture in the area for a few thousand years.

Koster people knew about early cultivated plants such as corn, squash, and beans long before they began to depend on agriculture as a way of life. Evidence from sites in the Koster locality shows that squash was in the area by 1200 B.C., corn by 200 B.C., and beans by A.D. 500. But the prehistoric peoples at Lowilva did not begin to depend on these cultivated plants until A.D. 900.

Several factors combined to delay the shift from a hunting-gathering culture to an agricultural one. First of all, the population growth rate at Koster was so gradual over the centuries that the numbers of people placed no strain on the available food resources. And secondly, because of their clever strategies for taking large harvests of seeds and freshwater fish, Koster people were able to provide themselves with yields equivalent to those that were later to be derived by cultivating crops. If a group of people is able to take plentiful amounts of staple foods from the environment with a minimum of labor, the kind of hard work needed to produce crops would have little appeal.

This equilibrium between people and the land apparently lasted for several thousand years, until the population began to grow and people were forced to find ways to increase their production of food. By A.D. 900 they began to practice large-scale agriculture to meet these needs.

The shift to agriculture brings certain cultural changes. Since it takes regular tending to produce crops, people become more settled and organize themselves into larger work groups. This in turn calls for more social rules and more complex social organization. At Koster we can see some of the effects of these important cultural changes.

In about 3900 B.C. the Helton village at Horizon 6 covered about five acres; we estimate the community numbered between 100 and 150 people. Helton people were hunter-gatherers. By A.D. 1000 the Jersey Bluff people had established by far the largest town we found at Koster, at Horizon 1. The settlement covered at least twenty-five acres, and possibly accommodated about one thousand people. And by that time the Jersey Bluff people began to cultivate the quantities of squash and beans.

There are other cultural changes that accompany the introduction of agriculture. When people become farmers, they change their relationship to the land. In most hunting-gathering economies, people reap a natural harvest from the land but do not alter the landscape. In an agricultural subsistence base, the key to productivity is to make investments of labor to alter the landscape. And when people invest their labor in the land, they develop proprietary feelings toward that land. What these feelings eventually led to in Lowilva we were to discover later.

Another result from our reconstruction of the food-gathering habits of the Koster Amerindians is that we have been able to gain a more realistic picture of what kind of people our prehistoric predecessors were. They were human beings who had developed an extremely sensitive, sophisticated knowledge of their environment. They were able to determine the relative value of the natural resources available to them, and to use that knowledge to select the foods that gave them the largest yields and the best nutrition for the least amount of labor. It was an ingenious adaptation and enabled them to develop a much more secure and stable way of life than we had previously suspected.

The Koster Amerindian's technology was on a much lower level than ours, and their total cultural development was much more simple than those of the majority of modern societies. Yet they were able to assess the finite set of environmental conditions in which they lived and to make a highly successful adaptation to it, something which we too strive to do.

12

North America's Earliest Permanent Houses

Say the words "prehistoric people" to yourself and very likely you'll visualize human beings in or near a cave. Now think of the words "American Indians" and one of the things that will come to mind will be a tepee.

To some degree these stereotypes of the aborigines' homes in North America are based on reality. Many prehistoric North Americans lived in caves; some Amerindians lived in tepees.

But as the first Europeans to arrive here found, there was a tremendous variety of cultural traits among the Amerindian tribes, and this extended to their forms of housing. Koster is showing us that these differences extended to types of dwellings in prehistoric times too.

Some of the Archaic people who lived in eastern North America at the same time Koster was occupied lived in cavelike rockshelters. At the Modoc Rock Shelter, Prairie du Rocher, Illinois, about ninety-five miles south of Koster, some distant neighbors of the Koster people made their homes in natural cuts in the rocky bluffs overlooking the Mississippi River. However, Koster is an open-air site. Although there are outcroppings of limestone on the face of the bluffs just behind the site, there are no openings in the rock that would have provided shelter for humans.

Fairly early in our excavation of Koster, when we still were digging isolated test squares in the cornfield, Gail called my attention to one square in which diggers were working through the dark, thick, debris-laden soil of Horizon 6 (3900 B.C.).

I climbed down into the square, using the knotted rope, and joined Gail. She pointed to a channel-basin metate* and next to it, a fire hearth.

"What do you think, Stuart?" asked Gail.

The metate was unusually large, measuring about ten inches wide and about eighteen inches long. Gail and I judged it to weigh about about forty pounds. It had been made out of a very big piece of crystalline rock, such as granite or diorite.

The intriguing thing about this particular metate was that it was rounded on the bottom, and someone had dug a small pit in the ground, placed the metate in the hole, and carefully wedged pieces of limestone in the ground under it to keep it from rocking as it was used.

The fire hearth next to the metate also had been constructed in a hole in the ground. Possibly the same "housewife" had dug the hole, about eighteen inches deep, and then stacked up three or four layers of limestone to form a cone-shaped pit. The diameter of the hearth at the top was about eighteen inches. Some of the limestone rocks had been burned a deep red, and sitting in the bottom of the hearth was a mass of charcoal, looking exactly like the charred materials that are left when you barbecue in an outdoor grill.

After carefully examining these two features, I said, "Gail, I wonder if you're not encountering an area over which there once was shelter of some kind, in which the woman ground her food and cooked. It certainly looks as if the person who used this was settled. I doubt that someone would carry around a forty-pound metate while moving frequently from place to place, or dig a pit and wedge a large grinding stone securely if she were just using it temporarily.

"And if that fire hearth had been outside and exposed to the elements, even for a short time, it wouldn't look like that. It would have layers of silt over the burned wood,

---

* The term *metate* (pronounced "may-TAH-tay"), borrowed from Mexico, refers to a concave or flat-topped stone on which one places foods such as seeds or nuts to be ground or pounded into flour by means of a smaller stone, the *mano*, held in the hand. A channel-basin metate is one with a bowl-like impression ground out in the top in which to place the food.

and the rock sides would have fallen it. Let's keep a close watch on this area as you dig."

A few days later, Gail reported that a strange-looking feature was emerging in that square.

Coming out from one wall was what appeared to have been a trench. But we were puzzled, because as the excavators dug down, the trench seemed to change shape and size. The long, dark stain in the earth would shift one way and then another. It would also get longer or shorter as they went down.

"I can't figure that one out at all," I admitted. "We'll just have to wait a bit to see what appears as you dig deeper."

And then one day toward the end of the 1970 digging season our mystery was partially solved. As I walked up the slope to the site on one of my daily visits, Gail approached, holding out her hand.

"What do you think these are, Stuart?" she asked. "We found them in Horizon 6, in the square which has the trenches, metate, and fire hearth all together."

She handed me two small orange-brown objects. I examined them carefully. They were rectangular, about one and a half to two inches long, with ragged edges. The material was porous. On one side were clearly visible random lines impressed into the surface, crisscrossing each other.

As I realized what they were, I was elated, and I called the students together for an on-the-spot lecture. As I talked, we passed the mysterious little objects from hand to hand.

"These are pieces of burned clay wall plaster, called daub," I explained. "The Helton people who lived at Horizon 6 apparently used this material on the walls of their houses to seal them from the weather. They probably took very thin twigs, wove them together, and then laced these between vertical posts placed upright in the ground. Into this wall core, called wattle work, they then embedded a clay covering. This kind of woven-stick wall core covered by a layer of clay is called wattle and daub.

"Here, the house probably burned down, and the clay was fired and hardened so that it lasted to the present time. This kind of construction, wattle and daub, has not been found before Jersey Bluff times, until now."

After the students had returned to work, I stood for several moments looking at the small objects in my hand, thinking how two such insignificant-looking bits of burned clay could be the bearers of such important news.

Wattle and daub used in construction will last for as long as thirty or forty years. It takes considerable effort to build a structure with walls of wattle and daub, and it seems unlikely that people would go to that much trouble and effort for a building they planned to use only briefly or even sporadically.

Archaeologists have assumed that it was not until the aborigines in eastern North America began to practice agriculture (around A.D. 900 in Lowilva) and could build up food surpluses that they became sedentary. From the evidence found prior to Koster, archaeologists' interpretation had been that hunter-gatherers had to move constantly from one temporary camp to another in order to procure a decent living while depending wholly on wild animals and plants.

(At the time, I assumed that the houses had been constructed of wattle and daub. Later, I realized there was insufficient evidence for this. Nonetheless, the size of the trenches indicated that people in Lowilva had become sedentary at least 4000 years earlier than archaeologists had assumed.)

Later when the computer analyzed the material from the trenches, it was able to solve our puzzle about the strange-looking features and to confirm that these were, indeed, among the earliest permanent houses in North America.

The computer could "see" what we had not been able to discern with our eyes. There had been multiple episodes of building in one place, which accounted for the shifting positions and lengths of the trenches. The Helton people at Horizon 6 initially had dug a trench, stood wooden posts in it, and then filled it with earth and rocks to support the posts for the walls of a house. Later, maybe when a family grew larger, they decided to expand the size of their dwelling and shifted the walls of the house outward.

When we took down the major block of the excavation, we found more trenches, which indicated that there had been five similar houses built in the Horizon 6 village. The houses were rectangular, and all of them had been expanded from their original size. The walls of each house were set parallel to each other, about eight feet apart, and were from twelve to fifteen feet long. Because we found post trenches only for the long walls, we think that the ends of each house may have been covered with materials that have deteriorated, leaving no clues. The Helton people might have used animal skins in the winter and woven grass mats in the summer as coverings for the open ends of the houses.

The Helton people cut terraces into the side of the hill to create a flat surface for their house floors. The floor of each house in Horizon 6 was dug about eighteen inches below what was then ground level, so that people would have had to step up to go outside, or down to enter a house.

We didn't reach the Horizon 8 village level at Koster until a couple of excavating seasons later. By that time I'd become immersed in fund-raising activities and made only an occasional visit to Koster. One evening in the dining hall Gail joined me long enough to suggest I come by the site the next day to examine some interesting-looking features at Horizon 8.

When I arrived at the site, I realized from the covert looks I was receiving from the crew that they were going to spring some surprise on me.

Gail and I descended, by wooden steps and ladder this time, into what had by now become an enormous hole in the ground. Pointing to the ground, Gail grinned and said, "What do you think?"

Carefully I examined the surface of Horizon 8. Yes, I could recognize trenches running along the ground, similar to those we had excavated at Horizon 6 but longer. More houses! This was especially interesting since the Horizon 8 village was more than a thousand years older than the Horizon 6 village.

Set at intervals within the darkened outlines of the trenches were a series of dark, round circles. There was no mistaking what they were—the organic stains that were all that remained of what had once been posts made from trees and sunk into the ground to form the walls of a structure.

The entire crew had stopped work, to watch my examination.

I looked up at them and smiled.

"Do you realize," I said, "that you have just discovered the earliest *permanent* houses yet found in North America? Earlier dwellings have been found, but the post holes were very small, indicating that they were for temporary shelters. These guys, back in 5000 B.C., were cutting down tree trunks measuring eight to ten inches in diameter for those posts. Nobody in his right mind would spend that kind of labor for a temporary shelter. These buildings were meant to last for years!"

The grins beaming back at me from all those sweaty faces were simply beautiful.

The dwellings in Horizon 8 were much larger than those we found in Horizon 6. Each structure was about twenty or twenty-five feet long, by about twelve or fifteen feet wide. Again there was no evidence for end walls. Probably these, too, had been covered either with woven grass mats or furs. We think members of an extended family or a clan may have lived together in each of the Horizon 8 houses.

To form the framework of a house, the Horizon 8 people dug foundations about two feet deep with sloping walls, and then set large posts in these. The posts were wedged with chunks of limestone to stabilize them. Posts were set about eight to ten feet apart, and there is no evidence for smaller posts having been set between these. The Horizon 8 people also cut terraces into the slope to set their house floors on level ground.

We also found some daub at Horizon 8.

Later on, when we learned from our scientists about Koster peoples' clever strategies for collecting sizable amounts of wild foods at one time, we realized it was this adaptation that enabled them to build permanent homes and establish long-term villages.

The building of a permanent home for one's family seems like such a simple act. Yet it was of extraordinary significance in the lives of these people. It indicates that, possibly for the first time in the thousands of years since people had first trekked across the barren wastes of the Bering Straits pursuing game and moved across and down the breadth and length of two continents, they had achieved enough mastery over their basic resources to assure economic stability. They had learned how to exploit their environment so well that they knew that they could stay put in one place and go out from home base and procure enough food to live in comfort, year in and year out, on a permanent basis.

In the course of excavating the house floors at Horizon 6 we had also uncovered some strange features, whose purpose eluded us for quite a while. These were indentations in the ground, which came out from the side of the slope. They were next to the houses and had been dug into the terraces on which the houses sat. Each indentation was about six feet long and eighteen inches wide. Their most striking characteristic was their shape; they had been cut into the ground in S-shaped forms, each of which resembled a wriggling snake. Of course, when found at about six feet below present ground level, they had been filled with earth.

We went on with our other work, but when we would get together for staff meetings or for a beer, the conversation invariably came around to these features. What could these strange S-shaped indentations have been used for?

Next time my friend Lew Binford arrived in Kampsville for his annual lectures to the NAP field school students, we visited Koster, as usual. Lew was especially interested in observing a group of students who were piece-plotting*

While we were there, I showed him the mysterious S-shaped indentations.

Lew dropped to the ground, on his knees, and, as I had done when Gail showed them to me, he ran his fingers along the S curves. Then he said, "I think the Helton people at Horizon 6 were having the same problems with rainwater on this hillside that you're having. They probably kept having to bail water out of the floors of their houses, especially since they had set the floors about eighteen inches below what was then ground level. I think these indentations were ditches to catch rainwater as it washed down the slope and divert it from coming into their homes. That's a damned clever adaptation, don't you think?"

---

* Piece-plotting is an excavation technique used when we want very fine control over an area of the site. In piece-plotting the excavator works with a trowel. Every time he or she hits an item that is one-quarter inch or larger, the excavator measures the exact location of the item before removing it from the ground, using strings placed at right angles to two sides of the square. These strings have been placed in relation to Hub Q, the primary reference point for plotting items on a vertical axis on the site. Hub Q usually is a pole stuck into the ground after the surface of the site has been carefully surveyed. A point on this pole is designated for everyone to use as the primary vertical reference point upon which all measurements taken at the site will be based. As you look over a site, you may spot small stakes in the ground to which a red flag or tape has been attached. The flag or tape serves as a secondary reference point, relative to Hub Q. After recording the position of the item, the digger places the object in a small coin envelope; the envelopes are then sent to the CID processing laboratory for further recording, and to the appropriate laboratories. We use piece-plotting rarely; it is too time-consuming and too expensive in terms of labor.

13

The "Kromebar People"

The year is about A.D. 20,000. Like the antihero in Ralph Ellison's *The Invisible Man*, you awaken to the terrifying realization that you are lying in a glass box. People dressed in strange attire peer down at you. As you blink your eyes and attempt to move, they are astonished. Within moments you have been lifted out and helped to stand, on shaking legs. The people around you are nearly hysterical, shouting at you and each other in a strange language.

Much later, after you have been examined by people who possibly are scientists or physicians and offered strange foods and drink, you try to communicate with your keepers to find out where you are and why. It takes awhile, in sign language, for you to grasp that you have been in some inexplicable state of suspension for more than 18,000 years, and that you are now on a strange planet, far distant from your own Earth. Slowly you begin to learn their language and to teach them yours. Among the many things that puzzle you is that they keep referring to you, among themselves, as something which sounds like "Kromebar."

Some time later you are taken back to the room in which you first returned to consciousness and, after a few moments of panic, realize they are not going to entomb you again. You recognize this is a museum, and, judging by the prominence which had been given the now-empty glass case which held you, you were once its primary display. Now your hosts invite you to examine the contents of other glass cases in the room, watching you intently as you do so.

The first case you encounter seems to contain about two dozen metal thimbles. In the next case, carefully arranged next to each other, are an aluminum colander, a small hand-sized electronic calculator minus its electric cord, the metal head of a garden rake, three hubcaps from a Ford, a gold wedding band, and a glass coffeepot which has been glued back together, apparently after being broken into many pieces. You realize you are being shown the material remnants of your own civilization. You stand before the next case, puzzled. It contains bowls of small metallic objects; it takes a few moments before you can identify these as pop-can tabs—thousands of them. After an initial wave of homesickness, you are baffled. Why are these objects arranged in such a senseless fashion?

Next, you are led to an adjoining room, in which, judging from your hosts' looks and gestures, something special is displayed. Here the glass cases are all the same size. In each case is displayed a six-foot-long gleaming metal object with bent ends. You are mystified. Why should these people display automobile bumpers?

One of your hosts points to the objects in the case and then to you and says, triumphantly, "Kromebar."

Having been an avid reader of archaeology books, it doesn't take you long to figure out that these people have recovered from your own extinct culture some artifacts which they have mistakenly assumed to be the most typical things made in your time. Apparently, when they were excavating, they chanced to dig up only that portion of an automobile plant in which car bumpers were manufactured. Following traditional archaeological custom, they have named your culture (and you) for the most typical artifacts recovered from it. That explains why they have been referring to you as one of the "Chrome Bar People."

As soon as you have learned enough of their language to make yourself understood, you explain their mistake and point out that they should refer to you as a member of the "Automobile People" instead. (Privately you sigh. You simply don't have enough mastery of their language to explain to them the subtle but important fact that back in 1977 some traditional archaeologists might have labeled Americans the "Automobile People" after some of their major artifacts, but that new archaeologists, who placed their emphasis on the study of the behavior of a culture rather than its artifacts, would have called them "Capitalists.")

Unfortunately I have no magic means for summoning back to life a Koster Amerindian from nine thousand years ago to explain to us what some of the mysterious objects we have dug up were used for, or how they were made. But if someone from the distant past could walk through many of our museums, I'm sure that person would be amused, or irritated, at some of our misconceptions about many of the artifacts on display.

All too often, archaeologists have guessed at how artifacts were made or used and have been content with their speculations. But if we wish to reconstruct the life-ways of ancient people accurately, we must use more precise methods to learn how they used their artifacts. And if we are to practice archaeology as a science, we must try to prove our hypotheses about the users of objects from the past.

I keep telling my students: The artifacts do *not* speak to us; we must figure out how to obtain important information from them. We have dug up literally thousands of items from Koster. The task of identifying the artifacts among those items, figuring out how they were made and used, and interpreting the behavior of Koster people from them falls to Dr. Thomas Genn Cook, director of the lithics (stone) artifacts laboratory in Kampsville and several other archaeologists involved in the Koster project.

Tom, long and lean, wears his dark brown wavy hair shoulder length, with a thick mustache to complement it. Dressed in jeans, plaid shirt, and cowboy boots, Tom resembles a tintype out of the Old West. But when he speaks, it is in the clipped accents of eastern New Jersey, and the phrasing is in academese. Now and then a sly wit flashes through and the dark brown eyes light up.

To identify the thousands of artifacts from Koster, Tom and his students go through the slow, laborious steps that are necessary to do formal and functional analysis of each artifact.

Formal analysis is the description of what the object looks like. For this, the observer lists a series of attributes, such as length, width, shape of the notches, color, weight, luster. If it's chipped, the chipping pattern is described.

Functional analysis is the description of those characteristics of an artifact which reflect what it may have been used for. If the piece looks like a pot, the observer would measure the volume, and describe the size and shape of the vessel's opening. In this way, one can distinguish a prehistoric water bottle from a pot used to cook stews. If it looks like an axe, the observer might include a description of the wear patterns or the shine on the edges that were used for chopping.

Each artifact is given a number, which, together with information to show the exact spot at which it was found, is marked on the object in India ink. These are called provenience numbers, and they are very important. When you are excavating a site, most of your action in digging is destructive. Once an artifact is out of the ground, if its context has not been recorded, then no matter how beautiful it may be, it is almost worthless to the archaeologist.

In identifying an artifact, the archaeologist always checks to see if a similar type has been found elsewhere. Tom examines collections of artifacts in museums or other laboratories, and he looks in publications on archaeology. If an artifact type is unknown, the discoverer gives it a name, usually labeling it for the owner of the property, or the area. When Greg Perino found a new type of projectile point while digging the Koster burial mounds, he named it the Koster point after Mary and Teed. We found some Koster points when we dug Horizon 1 at Koster, which let us know that the mounds Greg had dug were the cemeteries for the Jersey Bluff people who had lived at Koster.

The next step is to date the artifact by association. If charred materials have been found in the same occupation level in which the artifact was found, they can be dated by radiocarbon and this date in turn applied to the artifact. This process must be conducted with the strictest standards; the excavator must make careful observations and notes on where the artifact was found and must show proof that its context in relation to the carbonized material is unimpeachable. The digger must keep an alert eye out to make sure neither the carbonized material nor the artifact has been displaced from its original location, possibly by erosion, a rodent run, a root, or by humans.

If no charcoal is found at an occupation level, sometimes Tom can date the level from the artifacts themselves if they are similar to ones found at other sites that have already been identified and dated. The latter are then referred to as diagnostic artifacts.

In order to learn what people at Koster actually were doing, Tom analyzes the artifacts using two different methods. He studies the "tool kits"* which ancient people used, and he scrutinizes the artifacts in combination with the debris that resulted from their manufacture and use at the site.

At the beginning of his analysis of Koster artifacts, Tom came to me and said, "Stuart, there's no way we can do this job without some sort of help; there is simply too much material for us to handle, and too much information to be recorded and analyzed."

Jim Brown was planning to experiment with the computer for analyses of archaeological data as we dug Koster, and we decided to help Tom speed up his investigations with a computer program.

In addition, we came up with the concept of the central data-processing (CDP) laboratory, which occupies a two-story wooden building in Kampsville. A new coat of paint has helped to erase the weary look of the exterior, but the place still resembles one of the old saloons or hotels so familiar to viewers of screen westerns.

The morning after artifacts and debris arrive from Koster, they are washed. A familiar sight on a summer day in Kampsville is a long row of people standing at gray wooden tables in the yard next to the CDP laboratory washing artifacts and debris. The material is put on racks to dry in a neighboring building, where a battery of fans hastens the drying process.

Approximately two days after the material has come out of the ground, it has been washed and dried and is ready for processing.

Inside the CDP laboratory the debris is sorted into nine categories, including animal bone, hematite, limestone, flint, chert, and pottery sherds. Artifacts and debris from

---

* Archaeologists define a tool kit as an assemblage of tools used together in the process of completing a particular task. Consider, for instance, the kitchen in our society; all the items in that kitchen that are used in the preparation of food would be considered part of one tool kit. On a smaller scale, a plumber's or electrician's special set of tools also would be considered a tool kit. Usually, the more complex a culture is, the more numerous and varied its tool kits are.

each excavation unit are then weighed, by category. Specific types of artifacts are counted. All this information is entered into a log and later sent to the computer laboratory.

The CDP laboratory serves as an initial processing station for material retrieved from all the sites NAP digs. After being processed at CDP, animal bones are sent to the zoology laboratory; plant remains are sent to the botany laboratory; and stone artifacts are sent to the lithics artifacts laboratory.

This assembly-line processing of the vast quantities of artifacts and debris recovered at Koster and other sites is new in archaeology. Many archaeologists divide the excavation phase of their work on a site from the analysis of data and deal with the two phases separately. Consequently, at the end of the excavating season the archaeologist is faced with a sharp reduction in the size of the crew just when he or she needs help to wash and process large quantities of material. We find that digging and processing material simultaneously greatly increases our efficiency in conducting research.

Since Tom wishes to gain very detailed information about the material remains from Koster, he has his students "interview" individual artifacts. Tom has set up a series of twenty-four questions to be "asked" of each object, such as its size, what raw material it is made of, if it is made of chert whether or not it has been heat-treated,* whether it is whole or broken or unfinished. The laboratory works with six major categories of materials retrieved from Koster. These include chipped stone, ground stone, worked deer antler, worked animal bone, worked animal teeth, and worked mussel shell.†

At this writing, Tom's crew has made and placed on the computer file about 750,000 observations on 25,000 artifacts from Horizons 1 through 10. He is currently encoding data on artifacts from the rest of the horizons.

Tom tries to determine what tool kits were used by the Koster people. If a Helton villager at Koster Horizon 6

---

* The physical properties of chert are altered when it is heated.
† Chipped stone artifacts are made of chert, a form of flint, by removing flakes; ground stone artifacts are made from igneous rocks, such as granite or diorite, by being pecked, ground, or polished; "worked" means that an object has been altered in some way by human beings.

went out to hunt white-tailed deer, he would have carried a tool kit comprising killing and butchering tools. The killing tools were an *atlatl* (spear-thrower) and one or more spears. The spear included both shaft and stone projectile point. (The atlatl itself comprises four parts—handle, shaft, hook, and a shaft weight.) Among the butchering tools would have been a heavy cleaverlike tool for dismembering the carcass, a knife for cutting skin and tendons, a hammerstone for breaking open bones for marrow, and flake knives for slicing meat.

Once Tom has identified the various tool kits, he analyzes the spatial distribution of their parts in each horizon to learn where different human activities took place. Eventually he hopes to reconstruct the technology of each Koster village and to determine the relative importance of those technologies in the total lives of the people. He also wants to figure out how they structured their village to carry out specific tasks. He uses the computer to show where all of the artifacts were found in relation to each other and in relation to the various features on the site, such as the houses, fire hearths, storage pits, etc.

If a future archaeologist were to do this same type of analysis on a home which once stood in Evanston, Illinois, in 1977, he or she would find a tool kit for preparing food in one area; a tool kit for eating foods in another; a tool kit for writing, such as a typewriter, metal typewriter stand, desk lamp, and steel typist's chair, in a third; and a television set in a fourth. From the placement of these tools and the amount of space alloted to their uses the archaeologist would be able to figure out where the occupants once prepared food, ate their meals, did some kind of work involving typing, and relaxed, and might be able to gauge what relative importance each of these activities played in the lives of the extinct Evanstonians.

For example, judging from the relative thinness of the debris layers and the paucity of artifacts found at Koster Horizon 4 (2000 B.C.), in comparison to the thick, dark-stained layers of debris and immense quantities of artifacts found at Horizon 6 (3900-2800 B.C.), Horizon 8 (5000 B.C.), and Horizon 11 (6400 B.C.), it is apparent that people occupied Horizon 4 for a much shorter time and less intensely than the other levels.

When Tom analyzed the tool kits from the Titterington

people who had lived at Koster during Horizon 4 times,* he found mostly stone tools which would have been used for butchering animals, including knives, scrapers, and hammerstones. He concluded that in 2000 B.C. Koster had been used as a temporary bivouac by hunters. They apparently were away from their home village and used Koster as a camp for a few days or weeks while on a prolonged deer hunt.

Fred Hill's analyses of animal bones from Horizon 4 corroborated Tom's findings and threw some interesting light on Koster hunters' efficiency. Fred reported that the Titterington hunters were bringing back only certain bones of the deer to their temporary camp at Koster Horizon 4. The deer was one of the heaviest animals in their environment, and when they killed it, apparently the Titterington men dissected the carcass right in the field. Quite likely, they cut tenderloins from along the backbone and left the heavy vertebrae in the field along with the skull, from which they had extracted the brain. They cut the rest of the carcass into sections, bringing back with them only those portions that had the most meat on them, such as the haunches. By doing this, they considerably lightened the load they transported to their home base.

In contrast, at Koster Horizon 6, which is the densest occupation layer at the site, measuring almost five feet thick, we have retrieved several thousand artifacts. When we began to analyze the evidence from Horizon 6, we realized there had been three separate occupations there, succeeding one another over time.

After Tom had analyzed the tool-kit distribution from the top level of Horizon 6, he reported that people in that village did not bring raw materials for tools to the village but shaped chert into preforms elsewhere and then finished the tools at home. And his scrutiny also showed that this was, indeed, "home" to these people. Tom could distinguish tool kits for a variety of ordinary, everyday occupations, such as cooking, meat-cutting, making items from

* In 1938 archaeologist Paul F. Titterington excavated a Late Archaic period complex of burial mounds in the lower Illinois and Mississippi River valleys, immediately north of St. Louis. The culture layer later was named for him, and dated to about 2000 B.C. When Tom examined the projectile points from Koster Horizon 4, he found they were identical to those from the Titterington culture and applied that name to the people at Horizon 4.

grass or leather (perhaps clothing), and other tasks that would have been necessary to maintain a hunting-gathering way of life over several seasons of the year.

Horizon 6 people apparently liked to dress up, just as we do. They left behind copper and shell beads, which would have been used for personal adornment. They used red ocher (red iron ore, or hematite), mashing it into powder in little shell cups. Quite likely they decorated their bodies with red ocher powder for ceremonial occasions, just as many cultures today still do, including some Amerindians.

Tom also looks at tool kits from Koster for changes over time, to trace the changing life styles of the various groups which lived there over thousands of years. Although the Koster people may have been restricted to stone, animal bone, deer antler, and mussel shell as raw materials, they managed over the centuries greatly to expand and modify the types of tools they used, improving their technology. Apparently, as people became more settled and lived longer at one location, they did what most human beings do, they began to accumulate more things. And as they did, they elaborated their food-processing and food-procurement technologies, making each tool more specialized for its task so that they could perform the task more effectively and with greater ease. For example, a butchering tool kit from 2000 B.C. (Horizon 4) was more elaborate than one from 6000 B.C. There were more tools in it, and they were more specialized.

We were interested to find that as early as 6500 B.C., in the little hamlet at Horizon 11, people were using metates. This meant that they were already processing certain foods, such as nuts and seeds, for inclusion in their diet. The metates at Horizon 11 were rather small and shallow. Then in Horizon 8 they used grinding stones more closely resembling our mortars and pestles, which could hold more food at one time than the small metates of Horizon 11. By the time a larger village was established at Horizon 6, they had greatly improved their food-grinding tools and were using channel-basin metates (with a deep, teardrop-shaped depression in the center). This would suggest not only an improvement in the food-processing technique but possibly a greater use of seeds and nuts. So far, we have found no food-grinding implements in the occupation levels above Horizon 6, but we have found large metates dating to at least 2000 B.C. elsewhere in Lowilva.

As I mentioned earlier, beginning about A.D. 800 there were major changes in projectile-point styles which reflected an important alteration in hunting habits. After many thousands of years of using spears, first hand-held and later thrown with the aid of an atlatl, with stone projectile points ranging from two to six inches long, the hunters switched to bows and arrows, and their projectile points could now be made one to two inches long.

The introduction of bows and arrows provided the hunters with a much more efficient method of hunting and killing game.

Spears are heavy and the archaic hunter could carry only a few. The Woodland hunter could carry a pack filled with a dozen or more arrows. If the Archaic hunter was using an atlatl to launch a spear, it took a certain amount of time to reload. A Woodland hunter using bow and arrows could shoot rapidly and repeatedly, with little body motion to alarm the deer. An Archaic hunter with a spear could kill only one animal at a time. Woodland hunters, each equipped with bow and arrows, could ambush an entire deer herd and pepper it with arrows.

Sometimes we get a glimpse of ancient people that reveals how similar they were to modern humans in their quirks. Occasionally the Koster people turned out a tool in a new style that had nothing to do with how it worked. They did it to please an inner desire for something stylistically new and different, just as we trade in our automobile models or paint our living rooms a new color when the old car and the old color really are quite adequate. For example, we found stone axes at Koster Horizon 8 with grooves around them to which the wooden handle was attached. The Amerindians would split the wooden handle at one end, slip the axe into it so that the latter was held in the groove, then use wet rawhide or sinew to hold it firmly in place. When the rawhide or sinew dried, it shrank, binding the axe to the handle. People continued to use axes throughout the Koster occupations, but styles changed, although all were equally functional. By Horizon 2 times (200 B.C.) the axe was no longer grooved. Archaeologists call this type of axe a celt; it is attached to a handle by inserting it in a socket at one end of a wooden handle.

Tom also reconstructs the Koster people's behavior by analyzing the debris created when they made stone tools. Since we have retrieved literally tons of debris from Koster, we could never analyze it without the aid of the computer. And even with that marvelous superbrain to store and analyze instantaneously all the millions of bits of information on this debris, there is still an enormous amount of human labor involved.

With the computer-coded information about the artifacts and the many thousands of bits of debris, Tom attempts to reconstruct the series of steps that would have been used by prehistoric craftsmen to make raw materials into finished implements. The evidence he has are the stone tools and the debris resulting from every step in the manufacturing process. The problem is that he has all of it in his hands at one time, and all mixed together in one great mass of stone fragments.

When the CDP laboratory crew weighs chert, they do not subdivide it into categories which would reveal what stage of manufacturing each piece is from. For this Tom designated several classes of manufacturing debris and had the students go through the chert and sort it into categories. To our amazement, six of Tom's students were able to hand-sort and record observations on about eight tons of chert in one summer.

To set up the debris categories, Tom relied on his experience of the stages in making stone artifacts and observations he has made of other archaeologists turning out stone tools. For example, to make a projectile point, the artisan starts out with a block of chert, a large chunk of unaltered raw material; chips that down into a much smaller piece, called a *preform;* and then alters that to a finished artifact. In the process, he creates various classes of debris. When he first works on the large block of chert, he knocks off exterior flakes of very poor quality chert, which frequently results from frost damage. These are called *decortification flakes.* Next, the worker chips the remaining block down to the preform, a symmetrical form vaguely resembling the finished point. The flakes he chips off at this point are called *fracture.* Then, as he shapes the preform to a complete spear point, the pieces he knocks off are called *thinning flakes.* Tom had his students go through the debris and subdivide it into these and other categories and place information about each piece on computer coding cards.

When all these millions of bits of information about the debris categories have been stored on the computer, Tom can ask the machine special questions about manufacturing that can be answered from that collection of material. For instance, he might ask where an artisan was making projectile points in the Horizon 6 village. Does the evidence show he was making these from unaltered raw material, or that he was bringing in preforms that already had been cut from raw chert elsewhere? In Horizon 6 excavators found a scattering of chert chips near a fire hearth. Tom might ask whether there was a toolmaking station here or a tool resharpening workshop.

Tom uses this analysis to pinpoint precisely the specific behavior of ancient people, something archaeologists were unable to do in the past, without this type of analysis.

Incidentally, we find that the more rare a chert is, the fewer waste flakes we have of that material. Our inference is that these tools were made elsewhere and were traded to the Koster people.

To better understand what Tom is attempting to do by analyzing the bits of stone debris along with the artifacts from each horizon at Koster, let's assume that archaeologists in some future century happen to dig up our long-buried culture. (These are not the same people with whom you found yourself, at the beginning of this chapter, on another planet. Those people, you will recall, were not very thorough archaeologists and had mistakenly assumed that automobile bumpers were the major artifacts in your extinct civilization.) Let's assume that our society has been buried for many centuries. Wheeled vehicles have been obsolete for a long time, and people in this future society travel about by some as-yet-undreamed-of means.

Now suppose this archaeological team should happen to dig up Detroit (as it was in A.D. 1977) and they come upon the ruins of a factory that once turned out Chevrolet Novas. They would find, in the ruins, shells of several finished Novas, but the parts of each automobile that were made of perishable materials, such as seats (cloth) and the tires (rubber) would have deteriorated. The auto bodies, the machinery for making various parts of the auto such as doors, wheels, brakes, and windows, and the leftover debris that accumulated when these parts were being made would all be jumbled together in the ground.

These archaeologists, like Tom, would try to sort out all the thousands of bits and pieces of debris and to place them in categories so they could figure out when, in the manufacturing process, each was produced. Now they would not be going to all this trouble simply to find out how a Chevy Nova was made. They would pursue this line of investigation for what it could tell them about twentieth-century American life.

They would ask of their data: What were these objects, and what were they used for? How did they work? What raw materials went into making them? Where did the various raw materials come from? All societies must find raw materials to make their artifacts, and where they obtain these, and how, reveals a great deal about how enterprising they were, how far they traveled, whether or not they traded and with whom.

These archaeologists would observe that it took many people, working together, to produce these objects. And from the arrangement of certain areas of the auto plant, they could see that some people had been specialists, working on only parts of the auto, while others worked at putting together all the parts to turn out a finished auto. They would also conclude that our society had organized itself into huge, complex living centers in order to provide the number of workers needed in such a plant. And from the finished auto they would learn that we had manufactured an excellent mode of transportation. All of this would then lead to more questions: Where did twentieth-century Americans go in this form of transportation? Did an auto need a special track to travel on? What made it run? Who supervised all these workers? Where did they obtain their food? Who made the rules for this highly complex and technological society?

Tom has found that at Koster the manufacturing process (called a "tradition of manufacture" by archaeologists) varies through time from one prehistoric culture to another. Although the toolmakers might have gone through a similar series of steps to make a projectile point, different cultures went through these steps in a different order. So they produced different-shaped points and, in the process, different-shaped debris.

In the late Archaic period (2000 B.C., when Horizon 4 hunters used Koster for a temporary camp) they made large-core tools as an early step in the manufacture of projectile points, in which the artifacts were chipped over the entire surface, front and back.

Later, in about 100 B.C., the Middle Woodland people

in Lowilva had a "core-and-blade" tradition. They made special cone-shaped cores, and then removed from these unusually thin, sharp, parallel-sided flakes. This core-and-blade tradition produced the sharpest-edged stone tools ever created in the world.

Koster Amerindians may have been limited in their choice of raw materials from which to make tools, but they were quite ingenious with what was available. They had learned by 4000 B.C. at least that one could change the physical properties of chert by heating it.

Chert is a locally available material, the quality of which varies a great deal. Some of it is of very high quality, chips easily, and holds a fine edge in use, while other chert is granular and neither breaks well nor holds an edge. A person making a stone knife or projectile point would want the tool to have a very specific shape and size and to have a sharp edge. For this, he would choose a very high-quality chert, because there is less chance of a manufacturing failure. But if an Amerindian could not find good chert, he would use a lower-quality material and improve it by an annealing process called heat-treating. In heat-treating, the artisan takes the raw chert and roasts it in the fire to change its physical properties. The heat-treated chert is less brittle, and this allows the toolmaker to control the chipping more closely.

During Horizon 8 times, about 5000 B.C., the Koster people would first make a core, carefully heat-treat that core, then knock off very glassy flakes to make tools. But in Horizon 6 times, about 3500 B.C., they did it differently; they took large flakes and preforms, worked these to *bi-faces*, heat-treated those, and turned them into projectile points.

Later, in Horizon 4 times, about 2000 B.C., they didn't heat-treat the chert. Their tools were larger and look somewhat more clumsy. Apparently they had altered their tool sizes and shapes in order to make it possible to chip them from lower-grade material.

After working with the artifacts and debris for some time, Tom and his students have become so expert that they can identify specific manufacturing traditions that characterized a particular village at Koster. And sometimes they are able to identify the culture's chipped-stone-tool manufacturing process solely from debris from each step of the toolmaking process without having examples of the finished artifacts.

I mentioned earlier that new archaeologists believe that facts about the past are knowable, but must be proven, following the scientific method. Tom, as a new archaeologist, set out to test his hypotheses about what specific activities took place at Koster so that he could prove his conclusions to be correct. He reasoned that, as a person works, each act comprises three things. There is the task performed—toolmaking, butchering, cooking, or some other activity—the tool involved in performance of the task, and the resultant debris.

These three things—tasks, tools, and resultant trash—are linked together. For example, if an Amerindian were making ground stone items, he might have used chert hammers, hammerstones, and sandstone abraders. As he worked, he might have created trash that included unfinished ground stone forms, chert hammer fragments, perhaps some discarded tools, and pulverized stone.

Now if, in analyzing the Koster debris and artifacts, Tom found a collection from one spot that included chert hammers, hammerstones, and sandstone abraders, he could deduce that someone had been manufacturing ground stone items at that place. Not only could he deduce it, but he could *test* it by stating that the following types of debris would be found in association with the tools: chert hammer fragments, partially finished ground stone objects, and (if he were being fanatical about it) pulverized stone dust.

Before testing his hypotheses, Tom read accounts of historical and present-day hunter-gatherer societies, including the Naskapi, Ngulik, Cherokee, !Kung San, and Andaman Islanders, to see how they use their tools. From these observations, Tom designated three main categories of tasks which might have been performed by Koster people; extractive, maintenance, and social.

The term "extractive tasks" derives from the concept that technology is a means of extracting energy from the environment in the form of plant and animal products. An atlatl and spear would be used for extractive tasks—that is, for killing game. Other tools used for extractive tasks might have included a mussel shell to dig up duck potatoes and a stone blade to cut the roots of the duck potatoes free.

"Maintenance technology" is used to convert the raw materials into usable form. Metates and manos, knives, choppers, and bone needles all would be considered maintenance tools.

The term "social tasks" refers to the technology of social behavior, and archaeologists place in that category artifacts which were made and used for personal adornment or for ritual or symbolic purposes. Among these would be birdstones (polished stone artifacts in the shape of birds or dogs), shell beads, shell pendants, stone scepters, and items from a shaman's kit.

From the hunter-gatherer groups Tom came up with a list for his model of ten extractive and seven maintenance tasks that might have taken place at Koster, each requiring a specific tool kit. He didn't include social tasks in this model. For instance, Tom looked at the Helton culture at Horizon 6 (circa 3900–2800 B.C.), where excavators had found about 317 remnants of facilities, such as fire hearths, house floors, storage pits, etc. They also found more than five thousand artifacts and 600 kilograms (1,-300 pounds) of chert debris.

He was able to prove that the Helton people had engaged in hunting, fowling, fishing, and collecting nuts and other vegetables as food. Obviously these activities took place elsewhere than in the village (off stage, as it were), but from the tools made and/or used on the site Tom could observe the consequences of these activities when food and other raw materials were brought back for processing and/or consumption.

The Helton people also collected raw vegetable materials such as grasses and wove them into mats for baskets or cloth. They tanned hides from the animals they hunted to make leather items such as clothing or cooking pouches. And they made wooden items. There were no specimens of these kinds of artifacts, since they would have decayed, but Tom found indirect evidence of the manufacture of these kinds of items from the tools that were used. There were awls and needles, which would have been used to make leather or woven items. And there were stone tools resembling those used by historic Amerindians to work with wood. Some of the latter tools had wear patterns along the edge which confirmed their use for these kinds of tasks.

In the village proper, artisans turned out tools made of

chert (among them: projectile points, knives, and scrapers), ground stone (such as hammerstones to crack nuts or bones), shell (bowls or spoons), and bone and antler (awls, needles, or sharpeners). Helton craftsmen also spent a lot of energy keeping their stone tools in good working order, judging from the amount of resharpening flakes found in tool-working areas in the village.

People often ask me whether the Koster people produced art works. We have only found some beautifully proportioned bone pins, the handles of which are engraved in precise geometric patterns. They resemble small wooden clothespins, and we think they were used as hairpins or hair ornaments. Many have been found at burial sites, lying in the ground near the heads of both men and women.

The pins are four to six inches long, about one-quarter inch wide and one-quarter inch thick. to determine what raw material they were made from, Tom borrowed a deer skeleton from the zoology laboratory and checked the prehistoric bone pins against the modern deer bones. He found the bone pins were made from the metapodial (a bone in the lower leg) of a deer.

To make the bone pins, Koster artisans made incisions in the deer bone and split it in half to get a flat piece. Then they ground the flat piece and polished it with a grinding material we have found at Koster, which Tom calls "the original sandpaper." This is a lump of sandstone abrader; it is not found in the area. The nearest outcrops of sandstone are about 125 miles away. The final polishing on the bone pin would probably have been done with leather and a bit of dry earth to rub it down and get a nice, smooth surface.

Identical bone pins have been found at prehistoric sites in Indiana, Missouri, and Kentucky, an area encompassing almost half a million square miles. Apparently people were trading (possibly using the waterways as trade routes) over a vast portion of the continent as early as 5000 B.C. Most of the bone pins found at Koster were broken, and the debris found with them, in the form of chips of bone, suggests they were being manufactured at Koster by Helton people.

Tom's glimpse of the Helton culture, caught in one time frame from Koster, shows a settled people who made tools that enabled them to convert a wide range of raw materials into useful objects to help in hunting or gathering or to make life more comfortable at home. They were inventive; within the confines of a simple technology they turned out varieties of the same tools in different sizes and styles. And they produced tools that increased their efficiency at performing special tasks, such as bone awls for punching holes in leather, bone needles for sewing, tools for carving wood, and tiny stone drills, reminiscent of our smallest screwdrivers. They were creative and gave expression to a very human yearning for personal adornment by making handsomely engraved pins from deer bones. Apparently they had the leisure time to create objects purely for pleasure.

All of this evidence cited by our scientists may at times sound merely repetitious. From our point of view as archaeologists, that is as it should be. The more kinds of evidence we can accumulate that shows that certain activities took place thousands of years ago, the stronger our proofs. We try to analyze the behavior of the Koster people from many different aspects, so that we can reconstruct their life-ways as they actually were, not merely as we might assume them to have been. We don't want to make the mistake of labeling a group the "Kromebar People" when actually they may have been the "Capitalists" who manufactured automobiles.

When we choose a neighborhood in which to buy a home, we check it out for convenience to shops, transportation, and good schools. The Koster people apparently went through a comparable process of checking out their potential neighborhood for its advantages. One of the conveniences Koster offered was that it was located right next to two major sources of their most important raw material for making tools—chert.

People have used chert or flint (which are very hard materials, but easily fractured) to make cutting, shopping, and scraping tools for over two million years. There are only a few naturally occurring raw materials, such as chert, obsidian (a dark-colored volcanic rock resembling common bottle glass), and quartzite (an extremely compact granular rock), that are suitable for manufacturing sharp-edged instruments. Of these, only chert can be collected from nature almost anywhere on earth. Since people in prehistoric times were dependent on chert and similar rocks, it is vital for us to understand the methods and patterns through which chert sources were exploited, both for day-to-day work and for long-distance trade.

Thomas Meyers, an analytical chemist and archaeologist, is studying chert from Koster to determine where the aborigines obtained the material for their tools. He did a survey in a five-mile radius of the site and found that there were three particular types of chert available in the area. One type is found exposed in rock faces (such as in the bluffs right behind Koster), another is buried in clay banks, and the third is in stream beds.

Koster people made the majority of their stone tools from local chert; in fact, almost all of the chert they used came from the Koster creek, which ran a few yards from their homes. A small fraction of the chert in the Koster archaeological specimens is unlike that in the stream bed, and its closest likely sources are about two miles to the northwest.

In contrast, much later the Hopewell people in Lowilva made many of their beautiful artifacts from what archaeologists call "exotic" raw materials—that is, raw materials that were not from the immediate region. For a long time, archaeologists had speculated as to the source of the exotic raw materials found on the Hopewell sites.

In 1958 James B. Griffin, an eminent American archaeologist, and A. A. Gordus, a colleague in nuclear chemistry, used neutron-activation analysis to trace the origins of some of the Hopewellians' exotic materials. He was able to prove that all of the obsidian found in Hopewell sites in Illinois, Indiana, western Michigan, and Ohio came from one specific source of the material in Yellowstone National Park, in Wyoming, and an additional outcrop near Yellowstone. The Hopewellians, around the time of Christ, had engaged in trade over a vast part of North America.

The technique used by Griffin measures almost imperceptible quantities of various chemical elements in raw materials. The technique is so refined that it can detect and measure up to twenty-five chemical elements at once in a single piece of chert. Scientists call it "fingerprinting" a sample, because the eventual computer readout shows vertical markings of different lengths on a graph, and samples from the same geographical source will have similar markings, while those from different sources will be dissimilar. Neutron-activation analysis could be used to determine the trace-element composition of almost any artifact material.

Tom Meyers uses this technique to analyze chert samples from sites and chert sources in Lowilva.

We also use experimental-artifact analysis to enhance our understanding of how ancient people made and used tools. For this we teach our staff and students to make artifacts just as the aborigines did. In the process they can gain insights into Koster people's economy (How much chert did it take to make a projectile point?), working habits (How long did it take to make a point?), and level of expertise (How much skill and/or muscle did it take to make a point?).

One evening as I worked at Halsey House, the NAP headquarters in Kampsville, there came through the open window the tap, tap, tap of stone on stone. I glanced out, and saw a group of people sitting in a semicircle on the grass in back of the Kampsville Inn. The artifact-making class was in session, and I decided to drop by for a visit.

Moving through the group and occasionally stopping to direct one of them was a broad-chested man of medium height, whose luxuriant dark brown beard and bushy brown hair made him look like a biblical prophet. Sitting incongruously on top of the bush was a cream-colored ten-gallon hat. Guy Muto, NAP research archaeologist, was teaching the group how to make stone projectile points just like the ones Koster people made thousands of years ago, and just as he had learned from Don Crabtree, a specialist in stolen artifact experimentation from Idaho, one of the world's foremost experts on the techniques of chipping prehistoric stone artifact forms.

One of the things an archaeologist learns from replicating someone else's artifacts is never again to take for granted that the interpretation of data on a site is what it appears to be at first glance.

Pepe Hajic, a Northwestern University graduate student, was putting the finishing touches to a six-inch projectile point, banging away at the point with a granite cobble. She had started out with a four-pound lump of chert to make the spear point. Now most of the lump was debris, some of which lay at her feet.

"You know, Stuart, that suggests that for every artifact found on the site, there might be hundreds of pieces of debris lying about," said Pepe.

"Yes, the artifacts and debris found at a site may, in fact, reflect that no more than twenty people were living

there for a short period rather than five hundred people
for a long time, as you might first have assumed," I
replied.

We not only study tool making at Koster but also tool-
breaking. Wear on tools is a highly localized thing, and
since the activity for which a tool was used requires a specific movement, one should be able to see what kind of
movement is reflected in a particular pattern of wear.

For example, if you are using a sharp-edged stone in-
strument to chop with, you tend to get little hinge fractures along the edge, because the force is directed against
the edge vertically, causing little chips to pop off the edge.
But if you are sawing with it, you tend to get scratches
running parallel to the edge. So when you examine those
edges under a microscope, you can tell whether the edge
of a tool was used for chopping or slicing activities.

Guy's students plan to use the replicas they have made
of Koster tools to try to duplicate wear patterns that appear on artifacts from Koster. If they can duplicate the
wear patterns, they can further strengthen our interpretation of the use to which these tools were put.

Among the people who sat on the grass banging chert to
shape it into projectile points was Ann Koski, who is trying to figure out how tools were used by studying their
breakage patterns. Frequently we find tools at a site that
seem to have been broken at or near the same point in the
tool; there appears to be a pattern. Ann is experimenting
with placing different kinds of stress on replications of
stone tools to see if she can duplicate some of the breakage patterns we find. Again, the idea behind this research
is that using a tool as, say, a spear point would produce a
certain kind of stress, whereas using a similar object as a
knife would produce a completely different type of stress
resulting in a different kind of breakage. Consequently, the
patterns of breakage occurring in the points should show
how they were used and what kind of stress they sustained.

In another effort to learn something about Koster
people and their artifacts, our students make clay pots.
John White, an anthropologist who is part Cherokee,
teaches pottery-making and other Amerindian crafts to
students in NAP field schools. Under John's direction, the
students search out native clays within walking distance of
Koster or other sites. They make pots using techniques

similar to those that would have been used by the aborigines and turn out products resembling those made by the Black Sand people in about 200 B.C., when they lived in the Horizon 2 village. The students collect twigs to build a fire and bake their pots on this fire just as the Black Sand people would have done.

Recently John worked with field school students to make a canoe as the aborigines would have constructed one. They took the trunk of a freshly cut tree, designated one side the top, and built a fire on that side. They carefully nursed the fire until it had burned out a depression in the trunk to create the interior of the canoe. Then they finished the interior using a stone adze. The challenge, everyone agreed, was to get the fire to burn in a restricted area and at the proper temperature to achieve just the right amount of charring without burning through the tree.

I have suggested to Nancy and David Asch and Rose Duffield that for the next annual wild-foods hunt and picnic someone try to bring down a deer using a spear with a stone projectile point, or possibly with bow and arrow using a stone arrowhead. Perhaps Rose can cook her delicious wild-mushroom soup in a clay pot made by one of John White's students. And finally, maybe Ken Farnsworth or one of the students can go fishing in the Illinois River, sitting in John's canoe and using fishhooks made from deer bone like the ones Greg Perino has found at some Middle Woodland burial sites, all to enhance our insights into how Koster people made their living.

14

Keeping Track of Trash

We have all seen photographs in newspapers of trash left behind by crowds at a sporting event, rock concert, or beach. When you consider the amount of rubbish a group of human beings can create in only a few hours or a day, you can begin to appreciate the amounts of trash left behind by Koster people when they lived in a village for many years.

As new archaeologists, we aim to examine all that debris, along with the left-behind tools, house floors, fire hearths, and other features early people created, and to analyze all of those materials to learn what the people were doing at Koster. But this aim creates problems. The amount of material we collect is staggering. We could not begin to deal with it if we did not have the computer, with its extraordinary ability to remember not only a description of each item taken from the site but also the exact spot where it was found. (Of only slightly less importance is the computer's ability to remember where all of that material is stored in Kampsville or Evanston.)

The computer laboratory is housed in a small, gray former home on Highway 100 in Kampsville. NAP computer programs are directed by Dr. James Brown.

Jim thinks the best three dollars he ever spent was for a bottle of old Crow bourbon as a present for the supervisor of the University of Chicago computer operations, who let him and a friend use the computer for archaeological data analysis. That was back in 1961, when Jim was a graduate student at the University of Chicago. To earn money, he and a friend, Dr. Leslie Freeman, were doing some data analysis for Dr. Paul Martin of the Field Museum of Natural History of Chicago, one of the great senior archaeologists in North America, who was excavating a Pueblo ruin in the American Southwest. Jim and Les roomed with two sociologists who had used the computer in their own work. After watching the archaeologists struggle with their data, they suggested that they try the computer to speed up their work. So they introduced Jim and Les to the mysteries of the computer and set up a rough program. Many archaeologists were skeptical that the computer could be useful in archaeology, but when Jim and Les got results others became interested.

When we began to dig Koster and found ourselves excavating test squares that reached to thirty feet below ground, we realized we were going to have to retrieve and analyze enormous quantities of material. Jim suggested he try a pilot program using the computer to develop approaches to storage and swift retrieval of archaeological data. He received a grant from the National Science Foundation, which has supported the computer program for NAP ever since. NAP is one of a handful of archaeological programs in the world experimenting in a major way with computer analysis of data in the field.

The main purpose of the computer is to reconstruct each layer of the site on paper and to accurately locate artifacts, features, and debris in each horizon.

In order to reconstruct the specific activities that took place in the past, we need to examine the combinations of artifacts, features, and debris as they occur on a site. The genius of the computer is that it can tell us which combinations of objects were found on a site and where, and it can do this faster and more accurately than we can. Moreover, it can print out graphs called scattergrams and histograms. The scattergrams show differences in the distribution of objects in any horizontal plane in the site; the histograms show similar data in any vertical dimension on the site.

And, while in appearance the computer terminal is simply a set of typewriter keys attached to unseen computers hundreds of miles away rather than fashioned delightfully like a person (as in recent science fiction movies), it possesses "vision" which is far more acute than ours. It is able to distinguish subtleties in the combinations of objects on the site which would escape the human eye.

In addition to all this, the computer can perform instant analysis of material as it comes out of the ground and give us feedback on what we are finding, so that we can revise and improve excavating strategy. Deep sites like Koster pose tremendous problems because as we dig we do not know what we are going to uncover in buried occupation levels. On a surface site you can take a survey of artifacts, features, and debris and lay out the excavation to maximize samples of all these. But when you start going below the surface, in one sense you are digging blind, and what you find is largely dependent on where you happen to dig. You risk collecting too much of one kind of data and completely missing some material that may be important. You may end up with a biased view if you can't compare the centers of occupation levels.

Early in our days of excavating Koster I used to refer to the site as a "layer cake" of prehistory, and artists have depicted it that way. But the computer, which is able to gauge the contents of each occupation layer much more accurately than we can, has shown that when the site was being built up, layer upon layer, nothing so neat as a simple layer cake resulted. Some horizons are much thicker than others, and they thin out in different directions. For a mental picture of the Koster horizons, visualize a stack of pancakes tossed one on top of another, of varying thicknesses and sizes, with slightly different shaped edges.

This occurred because, when new people moved in, they did not establish the center of their village or hunting camp directly over the center of the previous occupations. Some groups lived more intensively at the east side of the site, others on the west or north portions of it. We don't really know how far each horizon reaches. We have dug test pits north of the main hole (the crew calls these the "back forty") to find the limits of Horizon 6. But we simply can't afford the number of hours it would take in human toil to do that for each horizon. We try to find the limits of the different horizons by drilling cores throughout the site, going very deep. Drilling cores is expensive too, but not nearly as expensive as it would be to extend the entire main block in different directions to cover each occupation to its farthest limit.

The computer alerts us if we are digging where a horizon is thinning out. The horizons at Koster sometimes dip, or change in soil color. When the computer prints a histogram showing the relative densities of material being encountered from top to bottom in a square, we can inform the site supervisor whether it is worthwhile or not to keep going down through that particular square.

We use the computer's magnificent memory for one more task—to retain the NAP Site Survey File data bank. Since we study the settlement systems of ancient people, we should ideally investigate all the sites at which the aborigines lived in Lowilva during a 12,000-year span of prehistory. Each summer we send out teams to locate and record archaeological sites within NAP's 3,200-square-mile research universe. So far, we have found eight hundred archaeological sites in Lowilva.

Much of our knowledge about potential sites comes from a network of about twenty-five local people, some lay archaeologists, some artifact-collectors such as Alec. Approximately eighty per cent of the land surface in Illinois has been plowed for agriculture. When a farmer plows, frequently he turns up artifacts that have been lying just below the surface. NAP's lay archaeologists and collectors constantly search for evidence of newly turned-up artifacts and report on them to us. These people include not only farmers but also a plumber, an electrician, a veterinarian's wife, and a mayor. All of them know their territory better than any of our staff; most have been collecting artifacts for years. Each keeps a watch on his or her own territory, and there seems to be an unspoken code of honor among them; one seldom if ever crosses into another's territory.

We store all the information on the artifacts and the sites in the computer. In the future, if we want to explore more parts of certain settlement systems, we can ask the computer to display information for us on the potential for such sites. For example, if we wanted to learn more about the Black Sand people, who lived in a village at Koster Horizon 2 (500–150 B.C.), we would ask the computer, "What is the evidence for Black Sand culture artifacts in Lowilva?"

We have had to build a special warehouse in Kampsville to store some of the massive quantities of artifacts and debris from Koster. After seven years, there's quite a bit of it. And still it comes. We have not yet finished taking down Horizon 11 and still have Horizon 12 and 13 to dig.

Sometimes, when I look at the floor-to-ceiling shelves in

the Richter shed (our original warehouse, long since filled to overflowing) and those of our new warehouse, I wince at the sight of all that material and the thought of the work it will take to make sense of it.

And then I think of our miracle-worker, the computer. Compared to science fiction models, it may seem stodgy. It can't walk or talk, climb sand dunes or pilot spaceships, but it can help us accomplish certain tasks which, in their own way, are quite fantastic.

One evening a group of us met for dinner at the Kampsville Inn. From the look on Jim Brown's face as he slid into his seat, I could see that he had something special to tell us. Jim normally is not an effusive person, but something in his manner had caught my eye as he entered the dining room.

He looked around the table and said, "The computer shows there were twenty-six distinct occupation levels at Koster, not twelve or thirteen."

There was silence for a few seconds as each of us struggled with the impact of this news, then the whole group began talking at once. It was a triumphant moment for Jim. Sometimes he has had difficulty trying to convince other archaeologists, and granters of funds, how important the computer can be in the analysis of archaeological data.

What Jim had just told us meant that people had lived at Koster on twenty-six different occasions over the centuries. We had been able to distinguish only thirteen horizons as we excavated. Without the computer, we would have gone along, assuming that was the number of different cultures that had lived at the site. Now, not only will we be able to study these cultures more accurately, but by having almost twice as many separate cultural units to study, we will be able to observe much more subtle and time-specific changes in human beings' adaptation at Koster.

15

Burial of the Dead

Imagine the puzzlement of archaeologists thousands of years from now as they try to figure out who was buried, obviously with very special treatment, by twentieth-century Americans in the Tomb of the Unknown Soldier in the Arlington National Cemetery.

They probably will be able to discern that this was a burial ground reserved for military people from clues such as brass buttons, brass and silver insignia, and bronze, silver, and gold medals found with the skeletons, although the cloth uniforms will have decayed. They also may find remnants of spent bullets or pieces of shrapnel among the skeletal remains, since these would not have been removed from a body before burial if the person died on the battlefield.

All of these clues, mingled together, may make it a bit difficult to sift out the relevant evidence. They'll ask: Was this site near a battlefield? And it may be difficult to determine exact dates if they find a soldier from World War I buried near a Civil War cannon and a tank from World War II.

No doubt they will recognize easily that some of the graves were those of high-ranking members of our society, because of the amount of space reserved for these burials and the monuments erected to them (such as the memorial to the late President John F. Kennedy).

But what will they make of the Tomb of the Unknown Soldier? Will they assume from the separate placement of the burials (there are three servicemen buried there—one each from World War I, World War II, and the Korean

War) and the amount of space around the graves, along with a large monument, that these people were also presidents, or generals, or admirals, or great leaders of some other rank? Assuming they have not yet learned how to read our language, how will they grasp the very subtle fact that the identity of each of these people, let alone his rank in the military or his status in society, was utterly unknown to us? And that that was among the reasons he was chosen to lie in this very special place? How will they be able to figure out that these burials, among all the thousands of others in the cemetery, were symbolic? And that we buried these men with special rituals and built a memorial to express our respect, appreciation, and grief for all of our lost and beloved war dead?

In every society the burial of the dead is done for the benefit of the living. In effect, it re-creates the social relationships of the living and the dead to each other. And frequently a cemetery reflects how the individual was ranked in the society, because that generally determines how the person is treated at death.

All cultures use various means to assign people to different roles and statuses. The two most common characteristics used are age and sex. (In American society, older men and women and younger boys and girls are listened to with the least respect. In corporate life, men occupy most of the executive positions; in the working force, most women hold menial jobs.) In almost all cemeteries, one is able to observe differential treatment of the dead according to how old the person was at death and whether the person was male or female.

Generally these cultural rules can be helpful to the archaeologist, but occasionally (as we saw in the imaginary case of future archaeologists at the Tomb of the Unknown Soldier) clues can be hard to assess correctly. This is why Dr. Jane Buikstra prefers to do her own digging. She is both an archaeologist and a biological anthropologist; she spends summers digging burial sites in Lowilva and winters analyzing the human skeletal remains she has dug up. She is one of only about ten biological anthropologists in the country who also dig sites.

Jane, working at a burial site, is hard to distinguish from one of her students. A pretty, shapely young woman, her usual field outfit consists of brief cut-off jeans, a cotton T-shirt, and a red-and-white dime-store bandana to hold back her shoulder-length curly brown hair. Her knees and sneakers usually are muddy, and from a back pocket of her jeans there protrude not one but two trowels.

Let me digress a bit now from the site itself. Koster is a habitation site. During most of the thirteen or more occupations there, people established villages. Like modern people, the aborigines for the most part buried their dead in special places away from their settlements. We have not yet found the main cemeteries for the Koster villages, with the exception of the one established by the Jersey Bluff people who lived in Horizon 1 (A.D. 450–650), which Gregory Perino had excavated up behind the Koster farmhouse in 1962 and had named the Koster mounds. He found about three hundred burials there. When we excavated the Jersey Bluff village in Horizon 1 in 1970, we found artifacts similar to the grave goods Greg had uncovered in the Koster Mound Group, and we realized that Greg had dug what was a cemetery for the Horizon 1 villagers.

We have found only twenty-three human skeletons in the Koster ruins, which—scattered through a time span of more than nine thousand years of prehistory—doesn't give a biological anthropologist much to work on. If analyzing these were all Jane had to do, she would have lots of time on her hands.

Jane chose to work in Lowilva because it offered her the opportunity to study the effect of the environment on prehistoric people as seen in their skeletons. People lived in Lowilva in prehistoric times back to at least ten thousand years ago. This implies that there must be a vast population of human skeletons, covering many generations, to be studied if we could find them. Part of the problem is easily solved. During the last 3,600 years of prehistory in Lowilva (from about 2000 B.C. to A.D. 1673) people buried their dead in easily recognizable mounds, many on the bluff crests that line both sides of the Illinois River, others on the floodplains near the river. (You will recall that it was a Hopewell burial mound, Kamp Mound Number 9, on the floodplain north of Kampsville that first attracted me to Lowilva.)

Some of the burial mounds in Lowilva are conical, built on the tops of the bluffs. Others are loaf-shaped, set in the floodplains. We were curious. Who built the cone-shaped mounds on the bluffs, and who built the loaf-shaped mounds near the river? Were they built by the same people, and if so, why in different shapes and sizes? Or were they built by different people? Were there any rules determining where each person was buried in these mounds, and how? From the exterior, all the cone-shaped mounds appeared to be similar and all the loaf-shaped mounds seemed to be alike. What about the interiors?

All of these questions appealed to Jane, and she decided to concentrate on two prehistoric cultural periods, the Middle Woodland (100 B.C.–A.D. 450) and the Late Woodland (A.D. 450–1200), because she suspected that it was during these periods that most of the mounds had been built.

But there were other reasons for her choice of those periods. I have referred to the fact that there were important cultural changes in Lowilva between the Middle and Late Woodland periods. Jane was intrigued at what the cemeteries and the human bones in them might reveal about those changes.

From our observations of Middle and Late Woodland period settlement systems we had learned that there had been a substantial population increase over time. The villages became larger. Some people were forced to move out of the main valley, where the most desirable wildfood resources grew in abundance, and to establish homes far up in the secondary valleys, where first-line wild foods were much less plentiful.

Along with the increase in the numbers of inhabitants in Lowilva, and their spreading out to settle over a greater part of the region, there was another extremely important development. In the latest phases of the Late Woodland, people began to cultivate beans, squash, and corn in quantity for the first time, though they continued hunting, fishing, and gathering for the major portion of their food. Possibly, as the population expanded, it became more difficult for each family to remain dependent entirely on wild foods as the total source of its diet. The change was gradual, but by A.D. 1000, people living in Lowilva had become heavily dependent on the crops they planted each year.

It was during the Middle Woodland period that the Hopewellians had flourished. They participated in a wide-ranging exchange network and imported exotic raw materials from many places in North America, including

copper from the Upper Great Lakes region, obsidian from
Yellowstone Park, and mica from the southern Appalachi-
ans. There also had been a flowering of artistic creation
during Hopewell times; they made artifacts that were
beautiful in design as well as useful. Among these were
clay figurines of humans and animals (which probably had
ritual significance) and "platform" smoking pipes (so
called because the bowl of each pipe was comprised of a
cylinder or animal effigy sitting on a flat platform). The
Hopewell people honored their dead by burying some of
their beautiful artifacts and sometimes pieces of exotic raw
materials with them.

For reasons we do not yet understand, the Hopewell
culture declined, and in the Late Woodland period the exchange network suddenly disappeared. Moreover, the very
beautiful designs they had created seemed to die out with
them. The artifacts produced by Late Woodland people
clearly were only utilitarian, with none of the playfulness
in design nor the handsome embellishments of the Hopewell objects.

So, as Jane set about digging Middle and Late Woodland burial sites, she was looking for clues that might help
explain why the population had grown, why the Hopewell
people had ceased to make their beautiful artifacts, and
what the effects were when people switched from total dependence on wild foods to living on cultivated foods, particularly corn.

Cultural and population changes are intimately related;
knowledge about one of these factors can increase our understanding of the other. One of the first things Jane did
was to send her students out to survey burial sites within
NAP's 3,200-square-mile research universe. She wanted to
know how many cemeteries there were, where they were,
how many people might be buried in each, and where the
cemeteries were positioned in relation to villages from the
same time period. Their observations confirmed that there
had been a population growth in Lowilva over time. They
also noted that the size of the population living on the
eastern side of the Illinois River (the Koster side) increased markedly over a long period in comparison to that
on the western side.

As she digs, Jane looks for hints of any forms of ritual
which may have been used at the time of death. For instance the American aborigines were known to anoint a

body with red ocher powder, either at death or during ceremonies surrounding death, and this is still a custom among some of today's Amerindian tribes. Some of our Euro-American religious groups have an analogous custom; at death they anoint the body with special oils, which have been blessed by a bishop. Jane also looks for grave goods, items buried with a body, which can tell her something about the social role and rank of the person with whom they were buried.

Generally, there are two kinds of grave goods found in prehistoric burials. One type is called contributed goods, which means they were given by the living to the dead, as a symbol of honor or respect for the dead person's status. For instance, in some societies there were obligations among clans; when a clan leader died, leaders from other clans contributed special goods to his grave. At Cahokia, near East St. Louis, Illinois, where the Mississippians (circa A.D. 900–1673) built the largest prehistoric human community in North America, high-ranking persons have been found buried with clusters of very beautiful artifacts. Artifacts in each cluster were of different styles, because they had been made and contributed by different clans or tribes.

The other type of grave goods usually are personal items, which were possessions of the deceased. Sometimes these were rare items which belonged to a high-ranking person and were symbols of his status or wealth.

If one were to compare grave goods from the tomb of King Tutankhamen with those from some of the graves of prehistoric North American aborigines, it would appear at first glance that they have little in common. King Tutankhamen's grave goods include many elaborately made items of materials which today are still precious. The items from the North American graves would appear relatively simple, although some were made of materials which were rare in their owners' societies. However, the two sets of items do have a great deal in common. For among King Tutankhamen's grave goods were mingled both personal possession and objects contributed by others to honor his great rank. And in North American aboriginal graves we find items which fall in both those classes.

Jane also can pick up clues to the person's status in the society by observing how the body was treated at death. One kind of information she records is the position of the body. Archaeologists use the terms "extended" and "flexed" to describe the positions in which skeletons are found. An extended skeleton is one which is found lying on its back, feet extended, arms at the sides. A flexed skeleton may be in almost any position which varies from that basic position.

In addition, the term "articulated" is used to describe the bones of a body when they are in the same position in which they occur in a living person. Hence, a skeleton may be in a flexed position, and at the same time it may be articulated—i.e., all of the bones are in the same position they would have been in at the time of death. Articulation or the lack of it reveals whether or not the bones were moved after flesh had decomposed or been removed.

Because she is so meticulous in her search for clues to social behavior, Jane has her students measure the degree of flexion at every joint. She has them record twenty-two different observations about the position of the skeleton on a form coded for the computer; they make these notations as they remove the bones from the ground. As part of their training, she teaches her students human osteology and gross anatomy. She has them dissect human cadavers in the Northwestern campus laboratory so that they will become familiar with the relationships of soft tissues to bones in the body. This enhances their interpretive powers in the field and enables them to recognize what the arrangement of the skeletal bones reflects about the way the body was treated at death.

Before they begin to dig a possible burial site, Jane's crew move across the area with a probe to test the soil in each mound for acidic content. A very low acidic content in the soil suggests there may be skeletons preserved there. Then they try to determine visually where the edge of the mound is and start a row of squares outside the mound in order to explore the natural stratigraphy, so that when they move into areas where people are buried, they can pick up the differences in soil color and texture.

Jane digs the entire burial site, including some of the ground around it, to make sure she is not missing important clues. She has found, for example, that sometimes there are also bodies buried in the flat surface next to a mound, as if these people had died after the mound had been filled with earlier burials. And in some Late Woodland cemeteries, where the custom was to cremate some people, she has found evidence suggesting there may have been charnel houses near some mounds. These were wooden structures (sometimes enclosed, sometimes an open platform set on upright logs) in which the dead were placed until the flesh had decayed; later, the bones were buried, sometimes after being cremated.

As the aborigines built their burial mounds, they went through several stages of construction. As she works, Jane peels off the material remnants of each of these stages in the reverse order from that in which they were originally put down. One group may have gone through as many as twenty-five different phases of activity to create a cemetery. Jane tries to distinguish each of these, just as we try to separate successive cultures in a habitation site. In addition, the burial population of any cemetery has been interred there at different times, usually over a span of years. Jane tries to date each phase she can delineate to get an estimate of the time span covered in creation of the burial mound.

Jane found that among the Middle Woodland (Hopewell) burial sites, the cone-shaped mounds on the bluff tops, and the loaf-shaped mounds in the floodplains had been constructed in similar fashion, with only slight (but important) variations.

For the bluff-crest mounds, the Hopewellians would choose a promontory overlooking the juncture of the main valley and a secondary valley. Right at the crest of the bluff, just before the slope began to descend, they would cut away the brush, clear the ground, scrape away the sod, and flatten out the site. There they would dig a shallow rectangular pit, about eight by ten feet, to a depth of about one and a half to two feet. Along the sides of the pit they would stack logs to form walls about four feet high, just like the walls of a log cabin. They used special clays as chinking between the logs, and then covered the log faces with woven mats. On the floor of the pit they would spread a thin layer of yellow sand. Finally, they would place a roof of logs over the open pit. This was to be the final resting place for their most honored members.

Next, they would build a circular mound of earth around the rectangular tomb, reaching to the top of the wooden crypt. Sometimes another, smaller mound was added, ringing the larger mound. The final effect would resemble a large earthen doughnut with a rectangular opening in the center; when there were two rings, it would look like a thick doughnut ringed by a thinner one.

The loaf-shaped mounds in the floodplains had larger central log crypts. One of the fascinating features about the mounds surrounding the log crypts was that they were made of soils of different colors and textures. At first it appeared that this was random and had occurred because burials were placed in the mounds at different times. But the pattern was repetitive and it became apparent that the use of different soils had some special meaning, which we haven't yet figured out. It probably was done as part of a ritual.

Many thousands of years from now, when archaeologists and biological anthropologists study our skeletal remains and social customs, they will have a much easier time of it than Jane does when she studies Middle and Late Woodland burial sites in Lowilva. (Unless, of course, all of our cemeteries, along with our civilization, have been blown into bits.) We usually bury our dead extended and articulated. Furthermore, in American cemeteries, we bury the dead in neat rows. But when Jane began to open the burial mounds in Lowilva, she found things were much more complex.

People were not buried in neat rows except in some portions of the mounds. And bodies had been treated in different ways at death before interment. Jane uses the term "burial programs" to describe the multiple steps used in burying people in a single culture as well as the many alternative burial requirements of different statuses. A future archaeologist, using Jane's terminology, would observe that our culture had several different burial programs. First of all, we bury our dead in different cemeteries, many of which are restricted to people of certain religious faiths. And we treat bodies in different ways at death: we inter some in the ground; we place some above ground in mausoleums; we cremate some, and either bury the remains, scatter them in the environment, or keep them in vessels in mausoleums or elsewhere; and we use some bodies for medical research. We also honor some people by interring their bodies in special places, such as Lincoln's Tomb in Springfield, Illinois, or Grant's Tomb in New York City.

We have other rules governing customs in our cemeteries. In many cemeteries the most desirable and attractive areas are reserved for the most affluent people. And there are rules in some cemeteries about the size of memorials that can be erected, or the types of flowers or plants which may be placed at a grave, or whether or not there must be a metal grave lining.

Similarly, the Hopewell people had different burial programs for distinct segments of their population, different types of treatment of bodies at death, and rules about who was buried where within the cemeteries.

In the place of honor, the central, log-lined tomb (in mounds both on the bluff tops and in the floodplains), several bodies were placed, extended full-length on their backs. Other bodies were placed in the doughnut-shaped earthen mound around the rectangular tomb. Still other bodies were placed in the smaller auxiliary mounds or else in pits dug in the flat ground around the periphery of the mound. These burials in the auxiliary ring-shaped mounds and the pits usually were what archaeologists call "bundle" burials. A bundle burial is one in which the bones are mixed together instead of being laid out extended. It may be that originally these persons were buried extended, either in the log-lined tomb or in the major doughnut-shaped mound surrounding the log tomb, and, after their flesh had decomposed, were disinterred and reburied as bundle burials to make way for later burials.

In the Hopewell cone-shaped mounds on the bluff crests there were twice as many males as females among the bundle burials in the surrounding large and small ring-shaped mounds.

Sometimes one adult male would be buried in solitary splendor in the central log tomb. Among the rest of the males buried in the mound, age seems to have made no difference in where they were placed, except that all were adult.

Buried along with the honored males in the central log tombs were a small number of women, adolescents, and children. They probably were close relatives of the high-ranking men.

The presence of the children along with adults in the central log tombs suggests that rank was inherited. Status in the group and, by inference, economic resources may have been controlled by a kinship group or lineage which handed down its power from generation to generation.

There also were children among the bundle burials in

the surrounding earthen mounds. But women and children made up only twenty per cent of the total number of individuals buried in the bluff-crest mounds. In the combined Hopewell mounds women and children made up only forty per cent of the burial population. Jane assumes that there was another burial program for this portion of the population; they may have been buried in cemeteries we have not yet discovered.

The burial program for the Hopewell loaf-shaped mounds on the floodplains differed somewhat from that in the bluff-crest mounds. Buried in the central log tombs on the valley floor (some of which were larger than those in the mounds on the bluffs) were a handful of males and a few members of their families. There were very few people buried in the surrounding rings of earth. In constructing the ring-shaped earthen mounds to surround the log crypts, the Hopewellians had carefully carried many different types of soil to the cemetery.

It was apparent that the Hopewellians had put a great deal of physical labor into building the central log tombs and the surrounding earthen mounds to honor a very few people at death. They had buried their most important men, along with members of their families, in the loaf-shaped mounds on the valley floor. Others of high rank, who were just below these people on the social scale, were buried in the log crypts on the bluffs.

All the men buried in the central crypts had been paid further homage. All had grave goods which included beautifully designed artifacts and pieces of exotic, rare raw materials. Among the items found with these men, who apparently were the most powerful people in their society, were quartz crystal pendants, wolf jawbones, copper necklaces, earspools, platform pipes, cut mica fashioned into either animal figurines or mirrors, and marine shells made into containers of various sorts and decorated with handsome designs.

In all, there were fourteen classes of Hopewell objects found in Middle Woodland tombs; eleven classes of fine objects were found exclusively with men. The most common artifacts found with women were long, thin, flake knives made of chert. These were found more frequently with women buried in the secondary mounds, the smaller doughnut-shaped mounds on the periphery of the larger ones.

Jane is still in the process of digging Late Woodland mounds and analyzing the data from these. There doesn't seem to have been much change in the design of the mounds from those in Middle Woodland times. Late Woodland cemeteries also contain central log tombs surrounded by earthen mounds.

But some of the rules for who was buried in the mound and how some of these bodies were processed seem to have been changed by Late Woodland times.

For one thing, in Late Woodland times many people were being cremated. Some bodies were placed on top of the mound, covered with bark, and cremated. Others appear to have been cremated elsewhere (perhaps after the body had been exposed for a while in nearby charnel houses). After cremation the burned bones were collected and interred as bundle burials in the surrounding circular mounds.

And apparently important changes had taken place in the rules for social status among the Late Woodland people. In Late Woodland mounds, adult females sometimes were given the place of honor reserved only for adult males in Middle Woodland log crypts. By this time, the society may have shifted from being a patriarchal one (where power and wealth are controlled through the male line) to a matriarchal one.

Another fascinating change shows up between Hopewell and Late Woodland grave goods, one which we have not yet been able to explain adequately in the culture as a whole. As in the villages of these two periods, the artifacts found in Late Woodland graves are not the beautifully designed objects found among the Hopewell people but rather are plain, utilitarian objects, usually made of materials found in Lowilva. Some archaeologists refer to the Late Woodland period as the "Dark Ages" in eastern North American prehistory because the very fertile creativity of the Hopewellians appears to have died out. Moreover, the lack of imported materials suggests that the far-flung trade network was no longer in operation.

We don't yet know why these important changes took place. Neither Jane nor I think there was less creativity. Possibly, with the growth of population, people were forced to turn their creative energies to finding a decent place to live and had to spend more effort figuring out how to find new food resources to feed their families.

They eventually did find these by turning from hunting and gathering to agriculture as their major source of foods. And as to the lessened trade, there may have been political instability caused by the greater competition for desirable places to live. All of this, however, is speculation. We will have to wait until we have more evidence to see whether any of these theories are correct.

16

There Were No Invaders

Each of us has in our skull many small holes, called *foramina* (singular *foramen*, pronounced "foh-RAY-men"), through which blood vessels pass. The number of such openings is determined genetically and varies among different people. Some of us have one hole above the right eyebrow, while others have two or more. Jane uses this and other genetic traits in human skeletons to trace migration patterns among Lowilva's prehistoric residents. Did foreign invaders sweep into the valley from time to time to overpower the natives?

She also uses genetic traits to study social customs among the aborigines. Did rich, powerful people marry only people from their own social class? Did people marry outside their clan? Was the Illinois River a geographical barrier to intermarriage between groups who lived on opposite sides?

Jane's detailed studies of human skeletons are innovative in archaeology. We still know very little biologically about our prehistoric predecessors in North America, because until fairly recently no one had studied their skeletal remains in any great detail.

First of all, the study of human skeletons had been focused on the question of the biological evolution of primates into modern human beings, and since most of that evolution occurred so long ago, in Africa and perhaps in part of Asia and still later in Europe, most of the research has been done there. Archaeologists in the New World have not found any extremely early human fossil bones comparable to those found by Louis, Mary, and Richard

Leakey in Africa. All the evidence so far suggests that by the time people crossed into the New World, perhaps earlier than 30,000 years ago, they already had evolved into *Homo sapiens*, which means they were physiologically the same as we are.

Secondly, many traditional archaeologists in North America who came across a burial site did not collect the skeletons, partly because an archaeologist working alone did not have the expertise to study human skeletons. Or the archaeologist might have collected a sample of skeletons from a cemetery and sent them to a physical anthropologist at a university or museum to be studied. But until the 1940s there just was not much interest in studying the remains of North American aborigines.

Some archaeologists collected only adult male skulls from a burial site and, at that, only a sample of skulls from each cemetery. They treated the skulls just as they did the artifacts they found, subjecting them to formal analysis. They would list the length, width, and height of a skull and describe some of its characteristics, such as heaviness of the brows. They would list the most typical characteristics seen in the skulls and use these just as they used a trait list of the most typical artifacts to describe a prehistoric culture.

Still other archaeologists took only skulls that were easily accessible. When they found a burial mound, they considered it primarily a place to search for artifacts. Some figured the fastest way to find out whether there were any artifacts in a mound was to sink a shaft in the center. They would then extract whatever artifacts could be found in this restricted area and take along any skulls that might be there, too, without bothering to look further in the mound.

As you might guess, this limited way of studying prehistoric skeletal remains led to rather biased conclusions about the cultures being studied. To give you an idea of what this kind of approach could produce, when earlier archaeologists examined some of the burial mounds in Lowilva, they were puzzled, as we were later on, by the variety of shapes, sizes, and locations of the mounds. When early archaeologists tried to explain this, they assumed that different-shaped mounds had been built by separate cultural groups and tried to explain the differences among prehistoric mound builders in terms of differences in skull shapes.

The flatheads, they decided, had built one kind of mounds and the roundheads had built another.

But when Jane conducted excavations in Lowilva, she found that the same people had built both conical mounds on the bluff tops and loaf-shaped mounds on the flood-plains during the Middle Woodland period.

Like the earlier archaeologists, Jane, too, found some skulls that were oddly shaped. But she saw another explanation. Hopewell mothers had bound their infants into cradleboards, which, depending on how it was done, could cause the infant's head to become misshapen. To further confuse archaeologists, customs in the use of these cradle-boards varied. If a mother placed the child so that its head pushed back hard against the board, the back of the child's head became broadened. But if she used a strap across the front of the infant's head, the forehead became flattened. Hence, the source of earlier archaeologists' designations of these cultures as "roundheads" and flat-heads.

There were other customs that could cause skulls to be misshapen. In the Middle Woodland period the aborigines usually buried a body face up, so skulls became flattened from the pressure of the earth on top. But in the Late Woodland period, bodies were buried lying on the right side, so these skulls tend toward a different type of warpage.

Analyzing the human skeletons from prehistory can help us answer a great many questions about the past. Did people live long, and what was considered a normal old age in Middle and Late Woodland societies? What was the rate of infant mortality? Did children grow at a normal rate? What diseases did people suffer from? Were they subject to the same diseases we are in our highly techno-logical society, or were some of their illnesses related to the kind of environment in which they lived? Did some people suffer from shortages of food? If so, what kinds? And what happened to their bodies when they switched from a high protein diet to a high carbohydrate (corn) diet?

Jane starts out by determining the age and sex of each skeleton. For different age groups, she uses different criteria. Modern studies have revealed that teeth in children develop at the same rate (it's fairly easy to predict that any six-year-old will be missing several "baby" teeth) and in the same patterns. Therefore, when Jane tries to establish the age of an individual who appears to be twelve years of age or younger, she examines the teeth. For older specimens, from adolescents on up, she uses other criteria. While people are growing, their bones do not grow at the ends but along the shafts. These bones fuse together, and do so at a predictable rate, so this measurement is one of the best indicators of age in a skeleton. For example, the distal end (the end of an arm or leg bone farthest from the trunk, contrasted to the proximal end, which is nearest the trunk) of the humerus (upper arm bone) fuses to the humeral shaft and becomes a single bone in females at about thirteen or fourteen years of age and in males at about seventeen or eighteen years of age.

After adolescence, it becomes more difficult to tell the age of a skeleton. As soft tissues age, so do the bones. The most reliable kind of changes in adults may be seen in the pubic symphysis (where the pelvic bones come together in front). There is a joint there which undergoes changes as the individual ages. The surface of the pubic symphysis becomes smoother and a rim forms around the joint; later this rim breaks down. Studies of these changes have been made among modern populations, where the age at death of each individual is known. Jane refers to these modern standards to determine the age of prehistoric skeletons, bearing in mind, of course, that individuals vary in the aging of soft tissues and bones.

She also looks at the skull, although this is not as accurate as the pubic symphysis. The skull in human beings is a series of interlocked bones, which fuse at a known rate, up to about age fifty-five. When Jane examines a prehistoric skull, she compares the sutures to standards that have been described for modern specimens and attempts to determine the age of the individual accordingly.

Jane uses several standard indicators to determine the sex of an individual skeleton. Certain changes in the pelvis and skull become prominent during adolescent growth spurts, and these can be used to determine the person's sex. In females, the pelvic outlet widens in order to facilitate childbirth. In males, the skull takes on certain more robust characteristics, at the ridges of the eyebrows and along the jaw, among other places.

We knew from the evidence at habitation sites that people had begun to plant corn in quantities by the end of Late Woodland times. But we didn't know what effect this had on their bones in comparison to the effects of the heavy protein diet (deer meat and hickory nuts) consumed by the Middle Woodland people.

If human beings are deprived of proper nutrients, their teeth will carry the evidence of it long after their death. Biological anthropologists look for signs of dietary stress in skeletons by examining the teeth. As children develop their permanent teeth, the enamel on each tooth is laid down in a series of bands, somewhat like a tree ring. If a person is not receiving the proper foods, the band of enamel laid down during that period will be less well formed than bands laid down while the person is healthy and well-fed. This is called dental hypoplasia. The inferior bands will be pitted; sometimes they can be seen with the naked eye. (During the Great Depression of the thirties many poor people in Appalachia suffered from malnutrition, and you can still see the effect of this on their teeth.) Jane examined teeth from both Middle and Late Woodland burial populations for signs of dental hypoplasia, both by looking at the outside of the teeth and by cutting the teeth into very thin sections and looking at these under a microscope.

First of all, she found that both groups of aborigines wore their teeth down much more rapidly than we do ours. Most likely this was because they consumed more grit and also because frequently they used their teeth as tools—for softening leather or as a third hand for holding and pulling. The teeth of Middle Woodland people showed marked wear in early adulthood, no doubt because of their rough diet, which included a lot of deer meat. The late Woodland peoples' teeth showed much less wear than the Middle Woodland peoples' had, but they had many more cavities. This would have resulted from their heavy carbohydrate diet of corn. And the teeth of Late Woodland infants also showed a great many cavities. Jane suspects these occurred when the babies were shifted from mothers' milk to a weaning diet possibly consisting of corn mush.

Neither the artifacts nor the human skeletal remains can speak to us when we dig a site, so we must use all our ingenuity to make sure our theories about the evidence are correct. Bioanthropologist Antoinette Brown did a related test on the Middle and Late Woodland burial populations, which confirmed our observations on their switch to a corn diet.

Antoinette studied the amount of strontium in the bones of the Middle and Late Woodland burial populations. Strontium is taken from the soil by plants. When a person consumes plants or animals that have eaten plants, strontium passes into the person's bones and is stored there. If the strontium has first been filtered by being passed through the body of a plant-eating animal, then less of it is passed on to the human being. Antoinette found larger traces of strontium in the bones of Late Woodland people than in those of Middle Woodland people. She concluded that the Middle Woodland people were eating more animal protein, and therefore the strontium was being filtered—some of it being retained in the bones of the animals before the animal protein was ingested by human beings. In comparison, Late Woodland people were consuming lots of corn and therefore taking in strontium more directly, as revealed by the heavier traces of it in their bones.

Jane is experimenting with studies of two other elements, copper and zinc, which can be measured in human bones and have great potential for telling us what kinds of food ancient people ate. Copper is an indicator of shellfish in the diet, while zinc, like strontium, gives clues as to how much corn a person consumed.

We can also learn whether prehistoric people were getting sufficient food by examining the bones of children and adolescents to see whether there were any interruptions in their growth rate. If children receive a nutritious diet, their bones grow at a certain rate. But if they suffer from poor nutrition, or from high fever or severe emotional stress, the growth of their bones may be interrupted. When growth resumes, concentrations of calcium build up on the bone, which show up as white lines on an X-ray film. These are called Harris lines, and they remain in the bone permanently, just as the lines from dental hypoplasia do in the teeth. To learn whether Middle and Late Woodland people had suffered from food shortages in childhood, Jane examined the long bones, arm and leg, of prehistoric skeletons by X-ray.

The bones of Hopewell people, from Middle Woodland times, showed Harris lines at regular intervals. This suggested that they had suffered from dietary stress from time to time, possibly in the late spring each year, when food might have become scarce.

Surprisingly, the bones from Late Woodland people showed no Harris lines at all. Did this mean that they ate well all of the time, had a fully nutritious diet, and never suffered from shortages? They were cultivating corn, and possibly they were able to produce enough food on an annual basis for everyone in the village. But Jane also took into consideration the fact that Harris lines are an indication of *resumption* of growth. Growth arrest lines occur in the bone as the result of episodes of distress; recovery from that stress is necessary for the Harris lines to form. Therefore, if a group suffers from chronic malnutrition, as do some people in areas of Central and South America and India today, then the lines do not form at all.

So Jane did further studies before formulating her conclusions. When she compared the length of the long bones of each individual with that person's dental age (which gave her clues as to the age of the person at death) and then compared the results from populations in both time periods, she found that Late Woodland people were attaining adult stature much later than the people in Middle Woodland times. They also had shorter lifespans and showed evidence of having suffered from more violent diseases.

Further work done by Dr. Della Cook, assistant professor of anthropology at Indiana University, analyzing these same prehistoric populations for signs of cortical thinning in bones also indicated that they suffered from malnutrition. (Della is married to Tom Cook. Like Jane, she is both an archaeologist and a biological anthropologist.)

Cortical bone is the dense outer layer of bone that supports the body, and it is a reservoir for calcium and minerals, reflecting activity or diet. People who are not well nourished lose cortical bone. A person who is not physically active also loses cortical bone. For example, if a person spends a month in bed, he or she loses a certain amount of cortical bone, and it's difficult to build it up again. The maintenance of cortical bone also depends on diet, and people whose diet is very high in animal protein tend to lose cortical bone because of the amount of phosphorus they take in.

Among juveniles from the Late Woodland burial populations, Della found much more narrow cortical bones than in juveniles from the Middle Woodland series. There also was a difference in cortical bone thickness in adults, with those of a given age in Late Woodland having less cortical bone than those in Middle Woodland series. Della thinks this shows that late in the Late Woodland period people were suffering from malnutrition as they were shifting their life from hunting-gathering to farming.

Della's observations on cortical thinning in bones corroborate Jane's findings that Late Woodland people had a more difficult time trying to wrest a living from their environment than Middle Woodland people did. We think that under the pressures of population growth, they were forced to more marginal areas and had to find new ways of obtaining sufficient food for their families. As they made the shift from a complete hunting-gathering way of life to that of agriculturists, they paid a heavy price physically.

Jane brings to her investigation of prehistoric diseases among Lowilva's prehistoric populations the same painstaking, step-by-step methodology she uses in her excavating. One day I was visiting her Evanston Laboratory and noticed a skull sitting on a worktable. The skull had a very neat hole next to the left eye socket, and I remarked, "Well, someone certainly took good aim at him."

"Yes, it does look at first glance as if he were killed by an arrow," replied Jane, picking up the skull. "However, if you look at the interior of the skull, you will see that the external hole is secondary to a lesion [sore] located within the braincase. You can see where the major focus of the infection was by that circle in the bone. See where the lesions appear to start in the region of the left eye? Possibly the primary infection started in the eye and then spread, probably rather rapidly, to the brain, where the inflammation of the meninges [meningitis] led to death. As the result of the eye infection, a portion of the bone around the eye became detached and was isolated; it didn't receive the nutrients necessary to maintain it, and it died. The general term for this isolated bone fragment is a sequestrum. So the hole next to the eye wasn't caused by an arrow but by a natural process resulting from the inflammation." The man had died in 500 B.C., yet Jane could analyze his fatal illness 2,500 years later.

Next, she went over and found a cardboard box on one

of the shelves, and brought it to the table. She took out several sets of vertebrae.

"Look, Stuart, they suffered from back trouble in early days in Lowilva, just as we do today," she said. "This poor guy, who lived almost a thousand years ago, suffered from an extreme example of a pathology that was common back then as it is now. Eight of his vertebrae were fused into an unmovable mass. A person suffering from this disease cannot flex or extend or rotate the spine. Today we call this affliction *ankylosing spondylitis*, or 'Marie Strumpell Disease.' "

Many aborigines in both Middle and Late Woodland times suffered from another common modern ailment—arthritis. But there were some fascinating differences in arthritic patterns among the men buried in Middle Woodland times, which suggests there may have been a division of labor between the elite and the less privileged. The high-ranking men who were buried in the central log tombs appear to have suffered from arthritis at the elbow joints, similar to that seen today in the elbows of baseball or football players. The males buried in places of less social importance, the earthen doughnut-shaped mounds around the log tombs, didn't have degenerative arthritic changes at the elbow. Instead, they had arthritis at the wrists, suggesting that they had been involved in some kind of activity involving the hands. This suggests that the high-ranking men were the hunters (throwing spears or using bows and arrows would have meant heavy use of the elbows), while the lower-ranking men were the craftsmen, possibly the toolmakers.

There were some other clues that suggested there might have been jobs that were reserved only for the privileged class. When Jane examined the skulls of Middle Woodland males, she found that the high-ranking individuals frequently had developed bony tumors in the cartilage of their ears (something which would have resulted in hearing problems). These tumors were not present in lower-ranking males buried in places of less prominence. Similar bony tumors appear today in the ears of people who spend a great deal of time swimming or diving, and Jane suspects that the high-ranking males may have played a special role in the procurement of aquatic resources for the community. For instance, it may have been their exclusive right to dive for deep-water mussels, or more exactly for the pearls to be found in such mussels.

Among the skeletal series from Late Woodland and later burial populations (the Mississippians, A.D. 900–1673) Jane has found evidence for what may have been an epidemic of a serious spinal disease in adolescents and young adults. Several of the skeletons—of both the young men and women—show lesions in the spinal vertebrae in the lower back. There are two modern diseases that produce lesions in the bone similar to the ones Jane has found in these prehistoric specimens—tuberculosis of the spine, and blastomycosis. The bony lesions in these two diseases are practically identical, but the rate of death in modern populations is different.

What is interesting about this finding is that blastomycosis results from a fungus picked up from the soil and occurs among people who work with soil. For farmers, it is an occupational risk. We suspect that these young people in Late Woodland and Mississippian times may have been afflicted with blastomycosis because they were spending a great deal of time cultivating plants. Jane's findings are not yet complete on this, but, if true, it would have been another severe penalty Late Woodland people had to pay as they shifted to agriculture as a way of life. And it would have been a contributing factor to shortening their lifespans compared to those of the Middle Woodland people.

As I mentioned, we have not yet found the main cemeteries for most of the people who lived at Koster, but we're confident that some day we will find them. Meanwhile, we have found the earliest prehistoric cemetery yet discovered in eastern North America, next to the ruins of the tiny hamlet at Koster Horizon 11 (circa 6500 B.C.).

The Horizon 11 cemetery was discovered one warm autumn day in 1975 when Tryge Widen's shovel struck a bone in the wall of the square he was digging at Horizon 11. Tryge secretly hoped it would be that of a black bear (the black bear has been regarded by Amerindians, from prehistoric times to the present, as an important and sacred symbol and has been treated as such in rituals), but it turned out to be a human skeleton.

Eventually Tryge and other crew members discovered seven more intact human skeletons in that spot, and the scattered bones of an eighth person. There may be more skeletons in adjoining areas of the site that we have not yet taken down.

Earlier, up in the "back forty," we had come upon another very small cemetery, with eight burials. This one dated to about 3900 B.C. and was created by the Horizon 6 people.

So far we have found twenty-three human skeletons at Koster. When one considers that the site was occupied many times over, during a span of more than nine thousand years, and that the majority of those occupations were villages in which people lived for long periods of time, then clearly the burials found so far can represent only a small proportion of the total burial populations. Since it was customary in later times, during the Middle and Late Woodland periods, to establish cemeteries fairly close to villages, we think this also may have been the practice among the Archaic people at Koster. Eventually, with some intensive exploration, we hope to find these buried cemeteries.

Jane has done a bit of preliminary analysis of the skeletons from the Horizon 6 cemetery to see what she could learn about Archaic peoples' burial customs. She wondered, after we had found the eight skeletons, why there were so few people buried next to what obviously was a large village. As we have seen, Middle and Late Woodland people had several different cemeteries for each village population, each of which was designated for people in certain social categories. Might the Archaic people also have had several different types of burial programs?

The only other excavation of a stratified Archaic site (and, therefore, one comparable to Koster) at which a series of human skeletons has been found is the Modoc Rock Shelter, located along the eastern edge of the Mississippi River Valley, at Prairie du Rocher, Illinois, below the confluence of the Mississippi and Illinois rivers. The Modoc Rock Shelter burials totaled twenty-eight individuals, who had lived there over a three-thousand-year period of prehistory broadly paralleling the Early and Middle Archaic periods at Koster.

When archaeologist Holm Neumann analyzed the Modoc skeletons, he was astonished at the high incidence of bone disease he found among them. There were numerous cases of striking bone deformity, including imperfectly healed fractures. and nearly every one of the people buried at Modoc Rock Shelter over that long time span had suffered from arthritis, Neumann assumed that the Modoc Rock Shelter skeletons were representative of the total population that had lived there. He thought that the high frequency of bone disease was a necessary result of the "rugged, difficult mode of subsistence" practiced by these hunter-gatherers.

When Jane examined the data on the Modoc Rock Shelter skeletons, she found that more than half of them were of people who had died when they were past forty. Since bone degeneration, due to normal aging processes, generally begins during the forties, it was not surprising that arthritis should appear in such high frequency in this sample. The young adults in the series all showed evidence of imperfectly healed fractures of limbs, which would have limited the range of activities these individuals could have performed. The only skeleton not afflicted was a youth under thirteen.

Then, when Jane examined the Koster skeletons from Horizon 6 (and one from Horizon 7), she found a remarkable similarity to the human skeletons from the Modoc Rock Shelter, The age-sex distribution roughly paralleled the Modoc group. Again, in the Koster group, most were adults well over forty who had suffered fractures or injuries that had not healed properly or severe arthritis, either of which would have limited their ability to perform many tasks. The rest were children under the age of thirteen.

All of the Koster Horizon 6 people had been buried with great care in the small cemetery and in similar fashion. All of the adults were male, buried on their backs with their legs flexed. All had their heads turned to the right. Young people were buried on their right sides. Artifacts were found with both old and young, including grinding stones, bone needles and punches, stone drills and projectile points, and antler tines (the points of deer antlers, used by historic Amerindians to chip spearheads from chert). Powdered red ocher had been either rubbed or spread on the bodies, indicating that it was used as part of the burial ceremony.

Jane concluded from both the Modoc and Koster skeletal series that during the Middle Archaic period (circa 5000–2500 B.C.) there apparently was a special cemetery reserved for those in the population who were not capable, because of either age (too young) or injury or chronic disease, of performing the full range of appropriate adult activities that would have been required in the society. At Koster Horizon 6, that special cemetery was placed at the edge of the village.

In 1969 Jane, together with Greg Perino, had excavated a series of burial sites across the Illinois River from Koster, just south of Kampsville, called the Gibson Mound Group. These were dated to the Middle and Late Woodland periods (100 B.C.–A.D. 1000). But in the lowest layer of one of these mounds they had found artifacts that differed in style from those in the later strata. Tom Cook identified them as being from the Middle Archaic period, broadly contemporaneous with Horizon 6 at Koster (3900–2800 B.C.). In other words, before the Middle and Late Woodland people had buried their dead in the mounds, earlier people, in the Archaic period, had used the same place as a cemetery.

So Jane examined the Gibson data to see whether they could give clues to any burial programs among Middle Archaic people different from those observed at Koster and Modoc. She found that most of the burial population at Gibson consisted of young and middle-aged adults, from about eighteen to forty years old, of both sexes, none of whom showed signs of striking pathology or arthritis. When Jane compared the Gibson and Koster skeletons, she found they were complementary in terms of age and sex distribution. The two cemeteries appeared to represent two alternatives chosen by Middle Archaic people as places to bury their dead.

So if you put together the evidence from all three cemeteries—Koster, Modoc, and Gibson—it becomes apparent that Middle Archaic people had, like the Middle Woodland people later, rules about where certain people were buried. People who were incapable of doing the work needed to keep the society functioning, either through injury or disease, and adolescents were relegated to a lower social rank (we know this from the burial goods interred with them) and buried in separate cemeteries. People from young adulthood to old age, who could share in the workload, were buried in places that ranked higher socially—the mounds on the bluff crests or terraces below the crests, and in stone crypts which were not mounded.

The fact that there were no infants or small children buried in either the Koster Horizon 6 cemetery or the Gibson mounds suggests that there was still a third track to the Middle Archaic burial programs, as yet undiscovered.

Jane also checked out the theories of earlier archaeologists, who thought it was not until about 2500 B.C. that the Amerindians had learned how to exploit wild-food resources efficiently. She looked at the Koster skeletons for signs of nutritional stress and found none. Middle Archaic people by 3900 B.C. were consuming very nutritious diets, indicating that they were efficient at making a living as hunter-gatherers.

As I mentioned earlier, Jane uses genetic traits and other physical characteristics in prehistoric human skeletons to learn about ancient people's migration patterns and social customs. When she examined the skeletons from Hopewell burial mounds, she discovered a fascinating fact. All of the adult males who had been buried in the most honored places (in the central log crypts, in mounds both in the floodplains and on the bluff crests) were taller than the rest of the population. They were the ones, you may remember, who also had been buried with the most elaborate grave goods. Who were these tall men? Were they perhaps invaders who came from elsewhere and subjugated the shorter local people?

Since stature often is an inherited characteristic, Jane set out to determine whether the tall men were of different genetic stock from the rest of the people in the same cemeteries. She examined the number of foramina in the skulls and other genetic traits. From her observations she concluded that the tall men and the rest of the population were, in fact, related to each other. There had been no invaders who had overpowered the Hopewell people.

Jane then wondered whether these men had received special treatment in their diets as youngsters. Suppose, for example, that during a time of food scarcity these males, as members of the highest ranks in the society, might have been given preferential treatment over children of lower-ranking people? If this had been the case, it could be checked out in an analysis of their bones, which would show they had undergone less dietary stress than others in the group. But X-rays for Harris lines and measurement of the strontium content in their bones showed no differences from the bones of the rest of the population.

People who intermarry within one clan or tribe often have children of shorter stature. One possible explanation could be that the tall Hopewell men were marrying outside of the immediate group. Jane speculates that since they were buried in places of honor, and apparently were leaders of the Hopewell, they may have been the ones who carried on trade with outside groups. In the course of dealing with other groups for trade activities, they may have chosen brides from outside their own social group and eventually bred more tall sons like themselves.

Cultures changed in Lowilva, and for this reason many traditional archaeologists theorize that from time to time invaders came, overpowered the local people, and brought about drastic changes. Yet look at our own culture. Eighty years is not very long, but in that time American culture has undergone incredible changes, since the introduction of the gasoline-powered engine and the first Ford production line. Will some future archaeologist, sifting the buried layers of our civilization, marvel at the changes in artifacts and features and debris from 1900 to 1980 in America, and speculate that the Automobile People came from somewhere else to overpower the much simpler Horse-and-Wagon People?

Did foreign invaders wipe out the Hopewell and, with them, their creative genius at turning out beautiful artifacts? Jane compared genetic traits among burial populations of the Hopewell and their successors in Lowilva, the Late Woodland people, to test this theory. As with the tall men among the Hopewell, she found there had been no invaders; the Late Woodland people were of the same genetic stock as the earlier Hopewellians.

About A.D. 900 a society known as the Mississippians developed in Lowilva. The Mississippians became heavily dependent on agriculture, particularly on corn. They also developed a highly complex society, with a priesthood as the ruling group. Eventually they built Cahokia, North America's largest prehistoric site, which covered more than four thousand acres. Among the Mississippians' artifacts were several which exhibited clear stylistic parallels with prehistoric artifacts from Central America and Mexico.

Again, archaeologists thought that the Mississippians had come into Lowilva, possibly from the south, and conquered local groups. But by comparing the genetic characteristics of Mississippian skeletons with those of skeletons from the Hopewell and Late Woodland burial populations, Jane found that the Mississippians were of the same genetic stock as other Lowilvans.

Later, when Jane examined the Koster skeletons for genetic traits, she found that there had been biological continuity in Lowilva for more than nine thousand years, from the people who settled the tiny hamlet at Horizon 11 to the Mississippians who founded North America's largest town.

Thus, by comparing tiny holes in the skulls of prehistoric skeletons (and other genetic traits), Jane found proofs to demolish thoroughly any theories about invaders. Her findings have very important implications for an understanding of the prehistory of eastern North America. There were no outsiders who came and imposed their ways on the natives. The cultural changes that took place stemmed from within the local groups, and it is there that we must look for explanations.

17

A 1,400-Year-Old Skirt

I had been out of town for a week, trying to raise funds for the expedition, and was rushing through the day, trying to catch up with the problems that had accumulated in my absence. As I came out of the Kampsville Museum, I saw Dr. Bonnie Whatley Styles standing on the porch of the zoology laboratory beckoning to me. She was smiling broadly and called out, "Stuart, you've just got to see what we've found!"

Bonnie is small and thin, with long, dark brown hair that hangs straight to frame a face with a porcelainlike complexion and sometimes covers the edge of the round, oversized tortoise-shell glasses she wears. Normally, Bonnie speaks in such soft tones one has to bend over to hear her; now she was practically shouting to me.

Bonnie is another of our NAP scientists who combines two research interests. She is an archaeologist-paleozoologist. Like Jane and Della, she prefers to do her own digging and then to analyze the animal bones she has excavated.

At the time, Bonnie was supervising a crew digging Newbridge, an early Late Woodland habitation site (circa A.D. 400–700) located right at the base of the bluffs on the eastern side of the Illinois River about eleven miles north of Koster.

As I joined Bonnie on the zoology laboratory porch, I wondered what kind of animal bones she had found to cause her to show such enthusiasm. She led me to the back of the laboratory to an old refrigerator in which the zoologists occasionally place a freshly killed animal or fish specimen prior to preparing the skeleton for the comparative collection. (The crew has had many a memorable feast from a large catfish or other specimen brought in, before adding the skeleton to the collection.)

"You're going to like this," Bonnie said as she took a large cardboard box out of the refrigerator and placed it on a nearby worktable.

Carefully she removed the lid of the box and then pulled apart layers of white tissue paper. There, lying on the paper, were several layers of black woven fabric.

I was impressed. I smiled at Bonnie, who grinned back at me and said, "Isn't it neat? I found it in the bottom of one of those pits at Newbridge. We figure it has to be at least fourteen hundred years old!"

This was a rare find in Lowilva, where, because of moisture both above and below ground, plant fibers, which were used by prehistoric people to weave cloth, are highly perishable. We knew that Middle Woodland (Hopewell) and Late Woodland people had woven cloth and mats from plant fibers. Sometimes we had found impressions in the ground in Hopewell and Late Woodland tombs which apparently had been made by woven materials that had long since deteriorated. Greg Perino had found very small fragments of cloth, and sometimes bits of leather, still clinging to copper axes placed in burial sites as grave goods. Apparently the axes had been wrapped in cloth or leather when they were placed in the grave, and since salts in copper act to keep destructive bacterial action down, a few fragments of cloth and leather had resisted decay.

But Bonnie's find was the first cloth we had recovered from a habitation site in Lowilva, and the first of any significant size.

The Newbridge site had been occupied during what archaeologists have designated as the White Hall phase of the Late Woodland period in Illinois. It has turned out to be a valuable site for data collection since the White Hall cultural layer proved to be pure and uncontaminated by other occupations. Frequently my colleagues and I have found that Late Woodland sites had previously been occupied by Middle Woodland people. All too often, there was not sufficient building up of sterile soil layers between the two occupations for us to sort out Late Woodland artifacts and debris from that of the previous occupants. As they dug at Newbridge, Bonnie and her students had encountered an unusually large number of features, which were densely concentrated.

"We were digging in six-foot squares, and were encountering two or three pits in each square," recalled Bonnie. "That's a lot of pits."

White Hall is famous for having round, flat-bottomed pits, dug several feet into the ground. The sides of the pits are very straight, and they are extremely neat on the bottom. Digging these pits is an amazing amount of work for us, with shovels and trowels, and the White Hall people dug them with mussel shells. Their patience must have been remarkable.

"I was troweling in a pit about three feet deep and thirty inches in circumference. [At NAP sites we always trowel a feature so as not to miss anything.] We could tell this was a feature at the very top of the pit, after we took out the plow zone, by the staining of the soil. It was very dark compared to the soil around it.

"It was obvious, after I began digging the pit, that people had reused the pit again and again. Up near the top there was a fire pit lined with slabs of limestone. Then about half way down I encountered a round clay liner, concave, which had also served as a fire pit.

"This second clay lining was important, I realized later, because it had actually sealed off the bottom part of the pit.

"And then, just below the clay liner, after I had removed it, I saw a little piece of black stuff. When I picked it up and examined it, I was just so excited.

"I ran over to the rest of the crew and said, 'You guys are not going to believe this!' "

"This" turned out to be a small fragment of woven black fabric about one inch square.

As Bonnie resumed digging, with the crew clustered about, she abandoned her trowel and worked with her hands.

"I was afraid that if I found any more cloth, my trowel might destroy it. It wasn't easy to dig in that pit, even with a modern shovel; at that depth we were running into a natural clay substratum that was hard to remove.

"I kept going down, and finally I hit an organic kind of fill, ashy, that smelled like a fresh fire pit. It was the strangest thing; it was like uncovering something that had not been buried very long. Yet later we were able to carbon-date it, and that fire had burned in A.D. 600, almost fourteen hundred years ago. When I thought back about it, I realized that the clay liner above had been the sealant, keeping those ashes fresh-smelling for centuries."

That wasn't all that the clay liner had helped preserve. Bonnie next encountered a neatly placed layer of grass stems; the layer was about two inches thick. The grasses were so well-preserved that they crackled when Bonnie touched them. And under the grasses, to one side of the pit, Bonnie suddenly saw a small pile of woven material that looked exactly like the small fragment of black fabric she had picked up in an earlier layer of the pit. The top piece, she estimated, measured about five by six inches.

Bonnie was ecstatic; she knew how rare it was to find preserved fabric in Lowilva. She leaned back on her heels and looked around at her crew, grinning and shaking her head in disbelief.

"We were afraid to touch it, Stuart, for fear it would crumble. I decided to leave it where it was, until I found out how to preserve cloth. I knew that if the fabric was dried out, it would turn to ashes and fall apart. So I carefully covered it with damp cloths."

On the day of the discovery Bonnie returned to Kampsville and asked among our archaeologists and scientists if anyone knew how to preserve fabric. No one did. So she called several museums, including two large ones in Chicago, to ask their advice on how to preserve prehistoric fabric.

"It seemed weird," Bonnie said. "No one could tell me how to preserve the fabric. They simply did not seem prepared for this kind of discovery. The Illinois State Museum curator told me to keep it damp, which I already was doing.

"Then I remembered having read an article by Dr. Margaret Brown [staff archaeologist in the State of Illinois Department of Conservation] in *American Antiquity*. She had described how to preserve bones and fabrics with a substance called polyethylene glycol and ethylcellulose.

"So I called Margaret, who was excavating Fort de Chartres [a historic site at Prairie du Rocher, Illinois,] and I drove down to see her. She gave me some of this liquid and suggested we experiment with it first. We thinned it down and tried some on the smaller fragments we had collected from the pit. There were a lot of broken pieces of fabric near the large pieces, and we'd taken some out and put them in the refrigerator.

"Back at the site, we took another look at the fabric and realized that not far beneath it there was another slab-lined fire pit. So we took that fire pit out as a unit, with the fabric on top of it. We removed the damp cloths and swabbed the fabric with the preservative. I didn't try to remove the dirt from the fabric in the field. The beauty of this preservative is that it is water-soluble, and I knew we could wash it out at a later date if we wanted to. So I just swabbed on this goo. Then we carefully reached underneath the fabric and lifted it up with a trowel and placed it in a cardboard box.

"When we got to the lab, we gently lifted the fabric out of the box, using spatulas and cardboard, to unfold it. We hadn't realized it until then, but we had four large pieces, each measuring about eight inches long by six inches wide. The edges of the cloth were frayed and looked as if they had rotted away.

"The Illinois State Museum advised keeping it cool, so that's why we're keeping it in the refrigerator.

"The large pieces were made with a very simple weave. The White Hall people—the women, no doubt—had taken pliable fibers and twisted them up into what I call bundles—I think that's the proper term—and wove them together using just a simple over-one, under-one pattern. The fabric is very thick and very tightly woven; no spaces can be seen through it. At the bottom of each piece the threads are loose as if it were a fringe. We thought this might have been a bag, and that the bottom had rotted out."

Bonnie asked John and Ele White to come look at the pieces of fabric since they are experts on Native American crafts. John noticed that the fabric had been more tightly spun at the bottom, and he suggested that this was an intentional fringe and that the pieces of fabric might have been part of one skirt.

Bonnie had noticed that, when she laid out the four large pieces of fabric, there was a lot of variation in the thickness of the bundles of thread. There would be thick bundles, even thin ones, and then again thick ones. Ele suggested that this might actually have been some sort of pattern in the cloth and that the cloth might have been dyed in different colors. This surprised Bonnie, who had thought of it as being black.

"It just never occurred to me that it might have been any other color," she said. "But, of course, it's all carbonized, that's why it's black. It was exposed to intense heat, which carbonized it, making it an inert substance and more resistant to decay. Fortunately, the heat wasn't strong enough to burn it completely and destroy it."

John also thought that the smaller fragments Bonnie had found might have been parts of a bag. These are made with space weft twining. There are Amerindian bags on display at the Field Museum of Natural History in Chicago made in the exact same style of weaving. We think this might have been a bag for collecting seeds.

"In the larger pieces, the long bundles of S-twisted fibers are the warps, and the fibers running through these are the wefts," Bonnie explained. "The wefts look as if they might possibly have been made from bark, and John thinks they may be made from basswood.

"You know, Stuart, this beautiful fabric may help to prove that there wasn't such a cultural decline during the Late Woodland period after all. I've always been a bit skeptical about the accepted theory that people in Lowilva plunged into the 'Dark Ages' after the Hopewell died out. It certainly took a creative person to produce fabric like this."

She ran a hand gently over a piece of the 1,400-year-old fabric and smiled, and I thought an archaeologist finding a king's tomb filled with gold objects couldn't have been more pleased.

18

Koster Rewrites History

The North American aborigines were not ignorant savages, as we have been led to believe by word passed down by Europeans who first encountered them in the New World. The Koster Amerindians and their neighbors in Lowilva were clever, intelligent people, capable of planning and executing complex strategies for taking abundant foods and raw materials from the wild resources in their environment to permit them to lead a very good life.

One of the benefits of the excavation at Koster and of the practice of the new archaeolgy in the New World generally is the provision of some firm facts about the Amerindians and about prehistory in this country, to replace the folklore which has been taken for granted for so long.

Many of us have learned about the aborigines (whom Christopher Columbus misnamed Indians) from the folklore of early pioneers. Historically, the whites who dealt with the aborigines saw them as obstacles to their own desire for the land, and therefore belittled the natives' achievements. Even when early whites opened prehistoric burial mounds which contained artifacts that compared artistically with those from sophisticated civilizations, they attributed the building of the mounds to some unknown race from another continent. They could not accept the truth, which was that the mounds, and the lovely artifacts, had been created by the ancestors of the aborigines whom they were displacing.

This contempt for the aborigines' culture is still held by many Americans today. In books and movies the Amerindians have been portrayed either as noble savages—ignorant, quaint, and romantic—or as barbarians. Both of these depictions are inaccurate.

Now archaeologists are piecing together the true history of the aborigines of North America, with some of the most important clues coming from the excavations at Koster.

To begin with, they were people just like us.

They were the ancestors of today's Amerindians. From her analyses of human skeletons found at Koster and in Middle and Late Woodland burial mounds, Jane determined that the people who had lived in prehistoric Lowilva had facial and skeletal characteristics similar to that of some modern Amerindians. In fact, dressed in jeans, T-shirt, and tennis shoes, and wearing hair cut shoulder-length, a Koster Amerindian could walk down State Street in Chicago today and go unnoticed.

When did people first arrive at Koster? Apparently, about 10,000 years ago. We have been able to establish a radiocarbon date for Horizon 13 at about 7500 B.C. There may be more cultural occupations below that level, predating Horizon 13. We know there were people in Lowilva as early as 9500 B.C. (almost twelve thousand years ago), because Clovis projectile points, which have been dated in the American Southwest and as far east as Massachusetts to that time, have been found near Koster.

As I review some of our findings from Koster, I'll be talking about Horizons 1 through 13 (with the exception of a few horizons which were so shallow as to yield almost no information). This may be a bit confusing to the reader, since earlier I mentioned that the computer showed there actually were twenty-six occupations at Koster. As we dug, we were able to see only thirteen horizons. Later, the computer was able to perceive that within those strata there had been an additional thirteen distinct occupations. For instance, where we saw one long-term occupation at Horizon 8, the computer could pick out four separate ones, each following closely on the previous one. At this point our scientists haven't finished their analyses of data from all twenty-six of those horizons. So I find it convenient, in presenting a look at the Koster findings, to refer only to the originally recognized thirteen occupations.

Up to now, archaeologists have assumed that in the period immediately following the retreat of the glacial ice, North America was occupied by peoples who were continuously on the move in search of wild plant and animal foods.

Yet here at Koster by 6400 B.C. Early Archaic people had settled down and established a tiny hamlet, about three-quarters of an acre large. About twenty-five people lived at Horizon 11 at the time, possibly in an extended family group. They settled there on a semipermanent basis, probably on an annual cycle, remaining during certain seasons of the year.

People occupied Horizon 11 over a very long time. The layer of debris from everyday living is about eighteen inches thick at its densest part; it thins down to about eight inches at the outer edges. Within that midden there are many fire hearths, some as large as thirty inches in diameter. A few were rimmed around the edges with chunks of limestone. The hearths had been used many times over to roast deer, mussels, and other foods. In some hearths, the burned earth is still bright orange, even after having been buried for more than nine thousand years. Near the hearths we found large metates and matching manos, the earliest evidence for food-grinding technology in North America. Women used these tool kits to grind nuts, seeds, and other vegetables. In fact in this area of the site, probably the center of activities for the group, we found large quantities of preserved food remains.

Traditionally, achaeologists have assumed that Archaic people went through a long, slow, gradual process in learning how to cope with their environment and how to extract a decent living from it. They thought it took the aborigines several thousand years, from Paleo-Indian times (circa 12,000–8000 B.C.) to 2500 B.C., to learn about various foods in eastern North America and how to exploit them.

This is simply not true. The Koster people knew their food resources intimately and did a superb job of feeding their communities. During the occupation of Horizon 11, Early Archaic people had developed a highly selective exploitation pattern of subsistence. They were not just taking foods randomly from the landscape. Rather, they calculated how to provide the community with the most nutritious foods possible while expending the least effort. In addition to deer and smaller mammals, they ate large quantities of fish, freshwater mussels, and nuts. Fish and nuts—in addition to being available each year, and easy to take in large quantities—are highly complementary components of a nutritious diet. Nuts contain fat for energy, which many freshwater fish lack. The kind of input-output analysis which was taking place was worthy of the most sophisticated culture.

We had still another surprise at Horizon 11—the discovery of two complete stone adzes,* tools that meant Early Archaic people were making wooden artifacts. An adze is considered a "secondary" woodworking tool. It is used for stripping, planing, or gouging wood. By contrast, an axe is a "primary" woodworking tool, used for chopping down trees and cutting up wood. Archaeologists had earlier found ground stone adzes dating back to 5000 BC. elsewhere. But now at Koster we found they were used as early as 6500 B.C.

The adzes at Horizon 11 are ground stone tools, made by polishing hard, crystalline rock, which would have been deposited in the area by glaciers. The adzes were probably made by being rubbed against other stones, pecked, and finally polished, probably with a piece of sandstone. Each adze had a chopping bit at one end and U-shaped gouging edge at the other.

The wooden artifacts that might have been made by using these adzes would long since have decayed. However, Tom Cook speculates that Horizon 11 people might have used the adzes to gouge out logs to make simple canoes.

We also unearthed evidence that strongly suggests these Early Archaic people were making baskets and leather items. We found bone awls (pointed instruments for piercing materials such as leather) and bone needles. The needles were made out of the long bones of deer, and one end had been honed to a sharp point. The awls were made from the ulna, or elbow joint of deer legs; one end had been blunted off to a round end, the other was pointed. Early whites reported seeing the Amerindians using these types of tools in their basket-making.

The stone adzes, metates, and bone awls, along with many different kinds of stone knives, scrapers, and projectile points, tell us this small band of people managed to turn out quite a variety of tools.

---

* Each whole tool is about 8 inches long, 3 inches wide, and 1¼ to 1½ inches thick. We also found parts of several others.

Two types of projectile points that we uncovered were similar to those found at other Early Archaic sites. One was a Graham Cave point, stylistically identical to those found at Graham Cave, Missouri. Another resembled a St. Albans point, found at the St. Albans site, Kanawha County, West Virginia. The dates associated with these projectile points at Graham Cave and St. Albans corroborate our 6500–6400 B.C. dates for Horizon 11.

Cultural ideas, customs, and styles radiate out from their place of origin as they are passed from one group to another. Was the Graham Cave projectile-point style first developed in Missouri and then passed along to Koster or vice versa? And how long did it take for such cultural traits to move from the place of origin to other points? These are still problems to be solved.

We also discovered at Horizon 11 the earliest cemetery that has been found in eastern North America. At the western edge of the little village three adults and an infant were buried. At the opposite side we located another adult male and an infant; and elsewhere at this level we found a third infant. The care that was taken in arranging these burials shows that the Early Archaic people had special rituals for the dead.

The dead had been placed in prepared oval pits, with their knees drawn up to their chests. The bodies had been left exposed until decomposition set in, for the soil had filtered in layers under and around the bones. In one case the dirt had filled the chest cavity. Some of the graves had been covered with large slabs of limestone or logs.

The body of one eighteen-month-old infant had been dusted with red ocher. This is the earliest known occurrence of this burial practice, which was common throughout North America down through historical times among the Amerindians.

Also on the western edge of the site, where the amount of debris from everyday living begins to thin out, we came upon three dog burials, all similar. (The first one had been discovered in 1970, by Nancy Wilmsen and Robert Kapka.) Each dog was interred in a very shallow pit. The dogs' legs were curled up, and the heads had been turned inward, toward the body. These burials must have had some ritual meaning to the Koster people. It is significant that this early the community was stable enough to domesticate dogs.

There is another bit of evidence from Koster that reveals a touch of human vanity—a few small beads made of hematite (iron ore). Since the nearest source of hematite was northern Michigan, apparently the Horizon 11 people were trading with others.

The next occupation, Horizon 10 (6000–5800 B.C.), tells the story of a completely different kind of settlement. At that time Koster was used only as a work camp where men came to make stone tools. The toolmakers couldn't have chosen a better spot; Koster was located right next to two major sources of chert—their most important raw material for making tools.

The toolmakers manufactured whole artifacts at their special camp, and also made artifact blanks. The latter were semifinished chunks of chert from which the worker later would form finished tools.

By Horizon 8 times (5600–5000 B.C.) people had once again established a village in the sheltered spot beneath the bluffs. Apparently the population had increased slightly, since the village now covered at least 1¾ acres. Actually, the computer reveals there were four distinct settlements during the occupation of Horizon 8, one on top of another. Horizon 8 is about two feet thick; one of the thickest occupation levels at Koster. We concluded that each of the settlements was a long-term one, possibly lasting for a hundred years or more.

During the first of these occupations Middle Archaic people built the earliest *permanent* houses yet found in North America. The size of the post molds indicates that Koster residents were cutting down rather large trees (ten inches in diameter), and it is highly unlikely that they would have expended so much labor for temporary shelters. They also invested time and energy to make their homes more comfortable. The builders cut terraces into the slope of the hill in order to have a flat surface on which to erect their houses.

We have other indications that Middle Archaic people lived in their village all during the year, from the kinds of plant and animal remains found there.

Having settled down to live in permanent homes on a year-round basis, the Koster people did what many of us do when we build houses of our own. They began to increase the number of their acquisitions. They added stone tools for cutting and shaving wood, called gravers, to their woodworking kits. Not only did they make more tools, but they began to improve them technologically. And in typically human fashion they also changed the styles of some of their tools. They added a groove to stone adzes and axes although it didn't necessarily make them any more efficient, just different in style.

After the Horizon 8 occupation, Koster appears to have been abandoned for slightly more than a thousand years, except for a brief occupation at Horizon 7. We have been unable to establish a radiocarbon date for Horizon 7, and the debris layer is very thin, so we can't tell much about it.

In about 3900 B.C. people once again had established a long-term village at Koster. Horizon 6 was occupied the longest of any of the Koster horizons; people lived there off and on for almost a thousand years. It also is the thickest horizon at the Koster, measuring about 4½ feet at its most dense. Again, the computer shows that there were three separate settlements in that horizon between 3900 and 2800 B.C. Judging from the tremendous amount of debris, these were all long-term occupations.

At this point the Amerindians introduced an extremely clever strategy in their exploitation of the fertile Lower Illinois River Valley. In fact, the Helton people who lived at Horizon 6 made one of the most impressive adaptations ever seen in a society that depended exclusively on wild foods and had only a stone technology. Although the Helton people were still hunter-gatherers, they had figured out a very efficient method for keeping their larders full the year around. They harvested annual "crops" of fish, ducks, geese, hickory nuts, and cereal-like seeds from such plants as marsh elder and goosefoot, which grew in dense stands in the Illinois River floodplains. This practice enabled them to gather in great quantities of food in yields which usually we associate only with agricultural production.

The community had grown in size, with about 100–150 people, and now covered about five acres. Possibly families banded together in larger social units to carry out the harvesting of naturally occurring "crops" of plants and animals in the valley and its environs. This kind of food-gathering technique would have been carried out most effectively by people working together in groups.

Some of the effects of this highly efficient adaptation can be seen in the fact that Koster people lived well into old age during the Middle Archaic period. Some of the human remains found in the little cemetery at the edge of the Helton village were those of people who had died at ages sixty, sixty-five, and seventy. And apparently, when the elderly became sick, they were well taken care of by younger people. Several of the people found in the cemetery were crippled with diseases of the aged, including severe arthritis, and would have needed help to move about.

You may recall that back in 1970 Richard Rawlins found the first human skeleton at Koster. It turned out to be that of an adult male who was well into his sixties when he died. He was buried in the Horizon 6 cemetery, but we did not know that when Rich discovered the remains, as we had not yet dug extensively at that level.

The people buried in the tiny cemetery at the edge of the Helton village represented only a small portion of the population. It was likely only one of three possible tracks which Middle Archaic people used in deciding where, and with what artifacts, a person was buried. In that small cemetery we found only people who, through age or injury, would have been incapable of functioning at full capacity. Most likely, the major portion of the population (including the able-bodied, productive people) are buried elsewhere, possibly in mounds on top of the bluffs. Whatever the criteria, we know from the Middle Archaic cemeteries which have been found that there were distinctions in social status. High-status individuals were buried with items of wealth made from rare materials and with symbols of rank such as bannerstones (specially shaped stones perforated with a hole, which were either worn on a leather thong around the neck or suspended from a stick, and which appear to have signified that the owner was rich or important).

It's not surprising that by this time Middle Archaic people had different social ranks. As one element of a culture becomes more complex, other parts usually do too. Not only did the Helton people improve their economic system by developing harvesting techniques, but they also participated in a wide-ranging trade network. Among the goods traded, they "bought" copper from up near Lake Superior, made beads out of it, and then traded these in exchange for raw materials, such as special flints which had been procured from quarries in southern Illinois and Indiana.

Horizon 6 people developed a more sophisticated technology for processing foods than had their predecessors. In the Helton village we found channel-basin metates—that is, ones in which a depression had been gouged out, to hold the food while pounding it, making them much more efficient than the previous, flat-topped metates.

It's interesting to observe that while the Helton people had discovered that clay, when exposed to heat, would harden and could be used for cooking, they did not take the next logical step and produce pots. They lined shallow pits in the ground with clay, then placed fire in these to harden them, producing clay hearth-liners, in which they cooked food directly with fire; they also may have placed hot stones in them together with food. It was not until three thousand years later that people began making portable clay pots in Lowilva.

After this sophisticated level of settlement and after three long occupations, Koster appears to have been abandoned again after the last Horizon 6 group lived there. We do not yet know why people left the sheltered spot under the bluffs after settling there over and over again. One of the reasons they may have left, recurrently, was because they had used up all the available firewood within a comfortable walking distance. When this happened, they probably moved off to another location. Later, when the forests near the Koster site once more had produced plenty of wood for burning, perhaps people passing by stopped, looked, and said, "Here's this beautiful spot. Why don't we move here?" Nancy's analysis of charcoal from the site showed that people at Koster were not fussy about what kind of wood they used for their fires, although some woods produce higher heat, and some burn better than others. They simply used what was available nearby.

However, it probably took a while for Koster to attract another large and permanent settlement. The next occupation, Horizon 4 (circa 2000 B.C.), was used only as a deer-butchering camp. Apparently the hunters lived elsewhere in the valley and camped briefly at Koster while out hunting. Before returning to their homes, they butchered the deer and left behind certain parts, such as the skulls and backbones, which would have been heavy to carry.

Most likely, they carved the deer up into manageable parts for transportation back home.

We can tell from our environmental studies that the climate at this time was becoming cooler and wetter. This meant a much richer source of game, and hunters particularly tracked turkeys, raccoons and deer.

From their projectile-point styles, we were able to identify the Horizon 4 hunters as members of the Titterington culture, a group which lived south of Koster near what is now St. Louis, Missouri. The Titterington people had complex burial practices and buried their high-ranking members with very beautiful artifacts, including pendants from marine shells imported from the Gulf of Mexico.

At Horizon 3 (1500–1200 B.C.) Koster was occupied by the Riverton people, a Late Archaic culture. This was a short-lived horizon with only a small amount of debris which told us little about these people. (We were able to identify this culture from projectile points they used similar to ones found at sites in the Lower Wabash River Valley, in Illinois.)

During Horizon 2 times people called the Black Sand were the first to produce true pottery at Koster. They left behind two types of pottery, Black Sand incised (meaning that designs were made in the wet clay with a sharp tool) and Liverpool cord-marked (meaning that designs were made on the pots by pressing reeds or straw into the wet clay before firing). We found no material for radiocarbon dating at Horizon 2, but similar pottery found at the Peisker site, eleven miles from Koster, led us to date this horizon to about 200–100 B.C.

The development of clay pottery marks an important technological improvement for Koster people. Until then they had cooked food either by roasting it over an open fire or by stone-boiling, placing hot stones, along with the food and water in either a woven basket, leather pouch, or clay hearth-liner. With clay pots the Black Sand people could now cook food over the fire without worrying too much about the degree of heat. And the clay pots provided them a means of carrying and storing food.

The most sweeping changes of all took place at Koster during the occupation of Horizon 1 (circa A.D. 400–1200). Technically speaking, Horizon 1 represents three horizons, for three different cultural groups lived there in rapid succession. In fact, they followed each other so closely that there was no time for a buildup of sterile soil from the bluffs. We found some of their debris intermingled, the only occurrence of this at Koster. However, we were able to distinguish pottery sherds and artifacts from three distinct cultures. The White Hall people lived at Koster from about A.D. 400 to 800; the Jersey Bluff people followed them, living there from about A.D. 800 to 1000. And then we found evidence that there had been an occupation by the Mississippian people, who lived there from about A.D. 900 to 1200. White Hall and Jersey Bluff cultures are included by archaeologists in the Late Woodland period; the Mississippian period (named for the Mississippian culture) overlapped the Late Woodland slightly.

The village during Horizon 1 times was more than five times the size of that at Horizon 6. The Jersey Bluff village at Horizon 1 covered more than twenty-five acres and as many as a thousand people may have lived there at the time.

During the occupation of Horizon 1, Koster people developed more effective hunting strategies as they shifted to the use of the bow and arrow, after having used the atlatl and spear for thousands of years. With an atlatl, a hunter usually stalked an individual deer, because he was carrying a heavy shaft which took time to reload. But with lighter-weight arrows and bows that could be reloaded rapidly, hunters could band together and conduct game drives, in which they herded a group of deer into one place and could then pepper the herd with many missiles.

But a much more important change took place—one that would have far-reaching consequences for all of the people in Lowilva. The Jersey Bluff people began to practice agriculture.* They cultivated squash and pumpkins in small gardens near their homes. The horticulture practiced at Koster was the precursor of full-scale agriculture, which Mississippian people later practiced elsewhere in the valley and upon which they eventually became dependent. The Mississippians at Horizon 1 were horticultural rather than agricultural. Like the Jersey Bluff people, they appear to have cultivated some plants, but on a limited basis only.

The shift from hunter-gatherers to agriculturists is very important, because it is one of the most crucial made by all

---

* Archaeologists use the term horticulture to imply small gardens, as opposed to agriculture, to mean the production of large crops.

cultures as they become more complex. Agriculture brings with it changes in the environmental, social, biological, and economic lives of the people. First they begin to change the natural landscape. They clear the land, chop down forests, or turn the prairie sod, and try to figure out ways of bringing water to the land. Not surprisingly, they develop territorial rights. Once the people of a village, or a particular culture, have put their energies into changing the face of the land, they begin to feel that this is their piece of land and that they alone have the right to harvest crops on it.

As hunter-gatherers, people lived and worked together in small groups. Perhaps the members of an extended family would co-operate as a unit. When they became agriculturists, they formed larger labor units. This, in turn, called for more complex social rules to govern who did the hunting, who did the farming, and how the foods gathered either way were divided.

For instance, among the Amerindians in historic times, some social groups were organized along a kinship principle, such as a lineage, where many families belonged to one lineage or clan. On a higher level or organization, there sometimes was a group called a moiety ("moiety" literally means half a group). The Cherokee Indians in the American Southeast were divided into two moieties, the Red and White; the men in one moiety were the warriors, those in the other were the farmers.

As they made the shift to agriculture, the Jersey Bluff and the Mississippian people also formed larger settlements. From about A.D. 950 on, we find the first large towns in Lowilva.

The shift from a diet of only wild game and plants to one that included cultivated plant foods, particularly corn, also affected the bodies of the Amerindians. Late Woodland people, the first to practice corn agriculture, not only had more cavities in their teeth than their predecessors but also a higher disease rate and a higher mortality rate in early adulthood.

Corn is a very storable product. Another effect of the shift to agriculture was that people in Lowilva were able to store additional surpluses of food to tide them over the lean winter months. Earlier, apparently, they stored dried meat and fish and nuts.

Prehistorians have spent a great deal of energy trying to

understand the role of cultivated plants in human history. This is a much more complicated story than we thought.

In examining the development of Western civilization, historians have observed how the cultivation of wheat and barley in the Near East, soon accompanied by irrigation systems, produced enormous increases in the yields of these foods over what the natural yields had been. In that part of the world, the shift to agriculture had a striking effect on the economic stability of the human population. Agriculture and irrigation offered many advantages over hunting-gathering as a way of life for the Near Eastern people.

Until fairly recently, very little has been known about how prehistoric people actually lived in North America. When scholars wrote about the prehistory of North America, they did one of the only things they could do—they looked at what archaeologists had learned about the early stages of Western civilization in the Mediterranean area, extrapolated some of the basic findings, and applied these in their reconstruction. From this, they conceived the idea that human beings inevitably go through a series of "revolutions" to improve their standard of living, among which is an agricultural revolution.

The Koster evidence suggests that human history and the involvement of human beings with cultivated plants is not always the same but, rather, varies with the character of the environment in which people are living.

Koster Amerindians knew about cultivated plants for almost two thousand years before they began to practice agriculture on any significant scale. We recall that cultivated pumpkin and bottle gourd were introduced in Lowilva about 1200 B.C., corn at 200 B.C., and beans at about A.D. 500. But Lowilvans did not begin producing any of these foods in quantities, nor to depend on them as the basis of subsistence until about A.D. 800–950.

Lowilvans did not shift to these new forms of food because their hunting-gathering-fishing way of life was so successful; they continued to practice it for several thousand more years. The Koster Amerindians had developed harvesting strategies for natural resources that gave them yields equivalent to those which later people derived from agriculture. By harvesting wild foods on a regular basis, they acquired a stability of life that anthropologists usually associate only with an agriculture way of life.

Perhaps one reason such stability has not been considered possible with a nonagricultural way of life is that historians have assumed that all patterns of human development followed that of Western civilization, which stems from the semidesert shoreline of the Mediterranean Sea, a place which contrasted greatly with the rich, fertile river floodplain of Lowilva in prehistoric times.

The Amerindians lived in Lowilva from at least twelve thousand years ago, yet only during the last one thousand of those years did they become farmers. We think that they turned to agriculture when the population grew to a point where the wild food sources were no longer sufficient. The river valleys of the American Middle West appear threadlike on a map; they represent an extremely small geographic area. Very high densities of natural resources were concentrated on very narrow strips of land. The number and kinds of resources in the surrounding upland forests and prairies offered a much lower productivity for people dependent on hunting-gathering as a livelihood.

Consequently, when the numbers of people began to increase, the amount of foods available in natural form must have become insufficient. In such a situation, people would have had two choices. Either they could have developed new social patterns and new economic systems that would enable the greater number of people to live successfully together and to be fed. Or part of the growing population would have had to move into more marginal areas.

During the early population increases which we have been able to trace, people chose the latter. They moved up into the secondary valleys, farther away from the densest concentrations of choice wild foods. There appear to have been three population growth spurts in prehistoric times in Lowilva. When Koster first was settled, in Early Archaic times prior to 7000 B.C., there seem to have been enough open areas in Lowilva for people to move into. The uplands, mouths of tributary valleys, and slopes at the edge of the main valley were occupied very early. The number of people at Koster increased only slightly for several centuries. The first real jump was in the Late Archaic period, circa 2500 B.C., during Horizon 4 times. At this time we find that the aborigines began to set up villages in the floodplains, closer to the Illinois River.

During the Middle Woodland period (circa 100

B.C.–A.D. 450) there was apparently another large increase in population. The number of settlements in the valley increased, and the villages covered a much larger area than those in earlier periods. This is the time when the Hopewell culture flourished.

Frequently, when we have excavated a Middle Woodland site in Lowilva, we have found a Late Woodland site on top of it. By about A.D. 800, Jersey Bluff Indians established very large towns, like the one at Koster Horizon 1, which contained about a thousand inhabitants. The population kept on growing after Jersey Bluff times, through Mississippian times until about A.D. 1400 at which time, mysteriously, population begins to decline.

At about the time that the last population spurt began in Lowilva, circa A.D. 800, the Amerindians began practicing agriculture to supplement their food supplies. They began to live together in larger social units, and we see the culmination of this trend in the river valleys of the Middle West in the establishment of North America's largest prehistoric community, Cahokia. In A.D. 1300 an estimated forty thousand Mississippians lived at Cahokia in an enormous city-state, which covered about five thousand acres. They were heavily dependent on agriculture for their food supply, and corn was their main crop. "Downtown" Cahokia, which occupied about seven hundred acres, was covered with earthen mounds. The largest of these, Monks' Mound, is 110 feet high—tall as a ten-story building—and covers sixteen acres. It is the largest earthen structure known to have been built by human beings.*

Many archaeologists see warfare as competition for scarce resources.

At about A.D. 800 the Jersey Bluff people were clearing areas of brush and planting crops. The first evidence for armed warfare in Lowilva appears at about that time.

When people begin establishing land tenure as an important aspect of agriculture, they are less flexible in re-

---

* Cahokia is a prime example of how Americans have ignored their prehistoric heritage. This precious site was 70% destroyed before archaeologists could study it. Most of it is covered by housing developments, shopping centers, and gasoline stations. Fortunately, in the first decade of the twentieth century, William McAdams, a lay archaeologist from Alton, Illinois, badgered the state legislature into setting aside something less than three hundred acres as a state park, and Monks' Mound stands on these.

sponse to neighboring pressures to move off into new collecting territory. This enhances the prospect for conflict.

Back in 1962, when Greg Perino excavated the Koster Mound Group, he found that a considerable number of skeletons had small arrowheads embedded in the bones or lying within the visceral areas or other parts of the body where there would have been soft tissues. He also found some skeletons with butchering marks at the cervical vertebrae, indicating that these people may have been beheaded. Jane also found some people in Late Woodland burial mounds who apparently had died from arrow wounds.

During the Mississippian period, people began to fortify their villages with wooden palisades. Cahokians built a great walled fortress, using wooden stakes around the central precincts of their city-state, enclosing about seven hundred acres.

If, indeed, warfare broke out at this time, the entire Koster site would have become uninhabitable, as the residents would have been vulnerable to attack from above. Attackers could easily have launched missiles from the heavily forested slopes and bluff crests next to the village. Koster probably would have become an indefensible location, and most likely this led to its final abandonment in about A.D. 1100–1200.

Koster poses some questions about the role of leisure time in the development of civilization. Most archaeologists have assumed that not until prehistoric people learned to domesticate plants and animals and to produce large surpluses of food did they have time to experiment with activities such as art, mathematics, and written language, which are characteristic of higher civilizations. Part of folklore is that people who live a technologically simple way of life are constantly involved in a search for survival. But in environments like Lowilva, where the wild food resources were dense, it was relatively easy to get enough to survive as long as the population remained constant.

There are not many hunter-gatherer groups—people who do not cultivate food but depend solely on naturally occurring animals and plants—left in the world. Some Eskimo groups still do this. Other modern hunter-gatherers include the !Kung San, who live in the middle of the Kalahari desert in Botswana, south central Africa; some of the aborigines in the western deserts of Australia. In all of these places, the environment is harsh. Modern populations have pushed these hunter-gatherers into areas that the more complex societies do not find attractive.

While hunting and gathering was much more bountiful for the Koster people in the lush environment of the Illinois valley, one can obtain clues to their life style by studying living hunter-gatherers even though they live in harsh environments today.

A group of Harvard University scientists, led by Richard Lee, has conducted an in-depth study of the !Kung San. !Kung San depend on game for one third of their diet. Among plant foods, they live primarily on the nut of the mongongo tree. This high-energy, high-protein nuts provides from half to two-thirds of their total diet.

Lee conducted statistics on how much work the !Kung San do during the worst period of the year, the dry season. In the !Kung San society, children and older people do not work; the latter retire at fifty. Nor do young people work until they marry; for females this is at about twenty, for males, about twenty-five. The people between twenty and fifty do all the work to provide for the population. They work an average of 2½ eight-hour days a week.

Since the Koster people left no written records, we have no direct knowledge of how much leisure time they had. But the statistics on the !Kung San would suggest that in hunter-gatherer societies such as those which existed at Koster, where food was abundant and easy to procure, the amount of actual time and labor needed to keep a small group of people eating well could not have been very great.

I think that it is not leisure time but competition that drives people to become more creative. Under competitive conditions, people develop those innovations and strategies, that insure them greater survival. Consequently, the people at Koster, while their population was small, probably enjoyed more leisure time than we do today. But when their population began to outstrip the available food supply, they had to develop new survival strategies, and agriculture was among them.

Since human beings first began to control animals and their breeding patterns and to cultivate plants, societies around the world have grown ever more complex. But until those changes occurred, the highest level of living had been achieved by the most complex hunting-gathering societies. When we compare the Koster hunter-gatherer cultures with any society that followed, we are struck by their simplicity. But if we look further back and compare the later Koster cultures with Paleo-Induan culture, then we can see that their adaptation to their environment was relatively complex.

Americans tend to equate success with size. Because North American archaeological sites do not yield monumental ruins like those found in Mexico, Peru, the Yucatan, and the Old World, many think that the aboriginal population of our country didn't accomplish very much. Almost any artifact recovered from Koster would fit into a shoebox except the metates and axes. Yet in terms of extracting a good living with minimum effort, the Koster Amerindians were extremely successful.

One of the goals of all human beings is survival, and one of the best measures of success is how well people survive. At Koster the Amerindians came and went from that one spot over a span of more than nine thousand years without any sign of cataclysm or replacement of local inhabitants through annihilation. Looking at genetic traits in human skeletal remains found in Lowilva, Jane Buikstra sees evidence for generations of biological continuity in the region.

These generations lived in equilibrium with their environment—something which we today are struggling to learn how to do.

They probably spent far less time in pursuit of a livelihood than we do today in our highly technological society with its many time- and labor-saving devices.

They made functional tools, which they kept improving and diversifying over the years. They were skilled at adapting many raw materials from nature for their use. They built permanent houses and were smart enough to protect them from slope wash by building very effective drainage ditches to divert water from their homes. They domesticated pets. They took care of their elderly, nursing them well into old age, and when their loved ones died, they buried them with special rites.

At times they participated in a vast trade network, which spanned large portions of the eastern, southern, and northern parts of the continent.

If one considers that twenty years is a generation, probably one of the more eloquent statements of the Koster Amerindians' success was their ability to endure in Lowilva for at least five hundred generations. (In 1976 modern Americans celebrated their own successful survival for ten generations.) The Amerindians might still have been living in Lowilva had it not been for the intrusive white population who moved into the area and claimed the land for their own purposes.

There are no monumental ruins at Koster, no elaborate artifacts. There is only the silent record, trampled into the earth by feet and covered by soft dust, wind-blown or washed down the slope of the bluffs by rains, over the centuries. In the ground are fragments of charred seeds, nutshells, pieces of animal and fish bones, mussel shells, and the tools of housewives, toolmakers, and hunters. In the small cemeteries people quietly sleep away the centuries. From these multitudinous fragments of evidence we are gaining an intimate glimpse of life as it was lived in one of the major river valleys of North America over a period of more than eight thousand years in prehistory. The picture which emerges gives a new perspective on America's first people.

19

A Day at Koster in 3500 B.C.

On a late fall day I sat on the bluffs just back of the Koster site, looking out over the deep hole and the Lower Illinois River Valley. It was a warm, sunny day, with a clear blue sky and soft white clouds. The leaves were in full color, and along the slopes of the bluffs the trees were brushed in soft, lustrous golds, reds, browns, and silvery grays.

After eighteen years I still thrill at the sight of this beautiful valley with its bright green floor (even on this late fall day) and the brown ribbon of river slowly winding its way south. I was tired and thought I'd relax a few moments before taking off in yet another airplane to address more groups of people in my never-ending effort to persuade Americans to contribute funds to help uncover their country's prehistoric past.

From below, a dog's bark rose, sharp and distinct.

"Gypsy!" Teed called his German shepherd, and I could hear man and dog as they went off together toward the hogpen.

The noise from a whole village full of dogs barking must have been ferocious in prehistoric times, I thought. The idea caught my imagination, and as I sat there, with the caress of the warm sun making my body feel blissfully relaxed, I found myself slipping into a fantasy of what the Helton village would have been like on a fall day in 3500 B.C., when Koster's Horizon 6 was occupied.

It was a few hours before dawn in the little settlement

that I could see in my mind's eye, and most of the villagers were still asleep, snug in their houses.

There would have been about 100 to 150 people living in the Helton village then. They were possibly divided socially into three or four clans of thirty to forty people each. The clan membership would have comprised a few extended families.

While it was still dark, men began to emerge from the houses and to gather silently in little knots, preparatory to going off for their day's work. As his master stepped out, a dog sprang up and barked, for which he was immediately rewarded with a kick. The other dogs moved about as quietly as the men, they knew what to expect if they behaved otherwise.

Several groups of young men, about six to eight in a group, each from a different clan, set out for the uplands, to hunt deer. The men, carrying atlatls and spears, ranged in age from about eighteen to forty, for deer-hunting required that a man be fleet and agile, in his prime.

Several other small groups of men set out to hunt ducks and geese in the backwater lakes in the river floodplains. There were young boys and older men along as well as young men, since this task did not demand quite the degree of spryness that deer-hunting required.

As they walked, some of the men chewed on pieces of jerky, dried deer meat. The women had prepared it by cutting the meat into strips and laying it out to dry in the air. Jerky (a version of which is carried by modern backpackers) makes excellent food for hunters or campers. It is light to carry, easy to store, requires no water or cooking, and keeps for a long time. Although it is stiff as a board, it softens as you chew it.

Had I actually been able to see these Helton huntsmen, they would have appeared very familiar to me, for they resembled some of today's Amerindians. They ranged in height from five feet to over six feet. We suspect they were dark-skinned, as their descendants are, and we know from a few skeletons that have been recovered with their hair intact that they had black, straight hair.

We don't know what the Helton people wore, but we assume they did have some form of clothing, because they lived in Lowilva all year round, and winters can be quite cold there. Also, the discovery of bone needles and bone punches make us believe that they worked leather. So I pictured the women in a simple garment of deer hide, sewn along the shoulders and maybe sewn down the sides or simply hanging loose, and the men in leather breech-clouts; later, as the weather turned colder, they too would wear some form of robe made of hides.

All was silent for a few hours, and then, as the first hint of dawn appeared over the eastern part of the valley, people began to stir. Stretching themselves, yawning, they came out of their homes and made their way to the creek right at the edge of the village for an early morning drink or to splash a bit of wake-up water on their faces. Some made their way to the area set aside, just outside the village limits, for their clan's toilet.

A few women carried baskets, tightly woven of root fibers, to the creek, where they filled them with water and carried them back to their homes. One woman helped her father, now in his seventies and badly crippled with arthritis, maneuver across the threshold of their home. He needed help, as he was barely able to move about on his own, and the floor of the house had been set about eighteen inches below ground level. The woman settled the old man on the ground near the house, where she knew he would catch the sun's first warming rays. Then she went to the creek to fetch some water. Finally she handed him a turtle-shellbowl containing some cold porridge.

There was no formal breakfast time in the village. In fact, there were no set mealtimes as we know them. The Helton people would eat at any time during the day when they were hungry.

For the break-of-day meal, whoever was hungry helped himself or herself to cold leftovers from the night before. The cold porridge which the woman handed the old man was made of mashed seeds from pigweed (*Amaranthus*) and marsh elder (*Iva*).

Today we take it for granted that people eat meals at regular times each day. The concept of regular mealtimes probably did not arise among prehistoric people in Lowilva until after they had been practicing agriculture for a while. In order to set regular mealtimes, the cook must know that food is available at predictable times and in predictable quantities. This is not possible when people are dependent totally on wild game and wild plants. The amounts of wild animals to be taken or crops of seeds, nuts, or greens to be collected may fluctuate, depending on the seasonal patterns of various animal species, the weather conditions affecting growing seasons of plants, and the ecological balance of insects, viruses, bacteria, and other factors that would affect either the animal or plant populations. Besides, hunters and gatherers may need to perform their tasks at totally different times of the day, according to when the resource is available or best taken.

The Helton people, like most hunter-gatherers, made a very sensible adaptation to suit the unpredictability of their food supplies and the varying schedules of their work parties. They practiced perpetual pot cookery. The fire was kept going most of the time, and the housewife maintained a continually simmering pot of food. Actually it wasn't until about 500 B.C. that clay pottery was first introduced to the Illinois valley, so the "pot" would have consisted of either a leather pouch or a tightly woven basket, into which hot stones were placed, together with food and maybe some water. The stew would include almost any type of food, such as game, fish, greens, root plants, some nutmeats, and seeds. Small fish were thrown in, bones and all; in cooking, the flesh would fall off the bones, and the bones would sink to the bottom of the "pot." To roast deer and other large game they would build large roasting pits, a bed of fist-sized limestone chunks over which a spit was erected.

Probably the only regularly scheduled meals in the Helton culture were ritual feasts, held to mark some special occasion, such as a very abundant harvest.

The sun was now up, and the women went about their housekeeping tasks. Some pounded acorns to make flour, after having washed the nuts through running water several times to leach out the acid. To make the flour, they placed a handful of shelled nuts on a metate and pounded and ground them with a mano.

Presently, from the western edge of the village there was the low sound of men's voices, and some of the children and the dogs ran to meet the returning duck-hunters. Women, like women everywhere whose husbands are about to walk through the door after a day's work, began to poke into their "pots." One woman stirred the contents with a long deer bone; a few added hot stones to heat up the stew; one tasted the food, scooping it out with a mussel shell.

When the men appeared, they were laughing and joking, and one quick glance revealed why. Slung over their shoulders were heavy strings of ducks, and an occasional goose or swan. Arrived home, they slung their catch to the ground near their houses and hastened to the fire to dry their wet clothing and warm up. A boy, seeing a dog nosing about the ducks he had just tossed to the ground, picked up a stick and tossed it casually toward the dog.

The Helton people had cast an eye skyward when locating their village. Lowilva is on the Mississippi flyway, along which each spring and fall thousands of wild waterfowl migrate today, as they have for thousands of years. The migrating birds are very set in their ways; they travel along a narrow corridor, without straying too far in either direction. As they pass through Lowilva, they stop off to rest and to eat. For the Helton people they were, in one sense, "free" calories since they didn't even take from the environment, except for a small amount of vegetation; they were reared in Canada and wintered in the South.

Still, it took knowledge and skill to trap them, and the Helton men had developed several strategies for this.

To catch waterfowl, the Helton men, after arriving at the edge of a large backwater lake, split up into parties of four. Each man stationed himself at one corner of a net, made of plant fibers and weighted with stone plummets.

In the darkness the men stood listening to the ducks. During the night the ducks sleep, resting on the surface of the water. Occasionally one duck will bump another and there is a brief gabble. The ducks could not see the hunters in the dark, and the latter were very careful not to make any noise.

Shortly before dawn, the men silently entered the water, carrying their nets. They waited until the last possible moment to avoid having to stand for too long in the cold water.

As the first rays of light began to glisten across the water, the hunters could just make out black lumps of ducks' bodies. They knew that any moment the ducks would begin to signal each other preparatory to taking flight. The men moved quickly, tossing each net over as wide an area as they could. As the stone plummets carried the nets down, the ducks were trapped, and the hunters clubbed them.

The Helton men may have used another method of capturing ducks. For this, they would build a cage called a weir, closed on three sides and the top, open on the fourth side. The cage would be placed in the water along one end of the marsh, usually where the water flowed out of the lake to form a natural creek. The cage would be about eight or ten feet high and about six feet wide. Leading up to the open mouth of the cage, the men would place a line of pickets to create a narrow corridor.

Just before dawn, the hunters would silently get into their canoes and, crouching very low, would gently and silently guide the boats toward the ducks. In the semidarkness, the ducks would perceive the canoes as logs and would be unwary. Gradually the "logs" would nudge the ducks toward the open end of the cage. As the canoes converged at the mouth of the cage, the hunters would paddle frantically, and the ducks, now alert and frightened, would fly up to the roof and sides of the cage, where waiting hunters would club them.

We have found no direct evidence for any boats at Koster, but some of the trade materials, such as special flints from Ohio, copper from the Upper Great Lakes region, Galena (lead ore) from northwestern Illinois or southern Missouri, and marine shells from the Gulf of Mexico, suggest that the early people were traveling by water. When we plotted the distribution of the origin of many of these raw materials, we found that many of the sources are right alongside navigable rivers.

Back at the village, some of the women began to clean the morning's catch of ducks and geese. Among the fowl the men had brought in were mallards, pintails, widgeons, and gadwalls. As they worked at skinning and cleaning the fowl, the women used several kinds of stone knives, choppers, and scrapers from their kitchen tool kits.

Each housewife kept her set of utensils in a corner of her house. The Helton houses were rectangular, about twelve to fifteen feet long and eight feet wide. The houses had been tucked behind the bluffs to keep them sheltered from cold winter winds. They sat scattered in a row along the creek.

Building the houses had been a communal affair. Each clan had worked together to erect houses for its members. The men had cut down trees to make posts that would form the support for their houses. Next, they dug trenches in the ground the length of the side walls and stood the wooden posts in these. Then they filled the trenches with earth and rocks to hold the posts upright.

Their houses were made of willow saplings collected in the Illinois River floodplain. The bark was stripped off and the saplings were bent and tied together with bark strips to form a dome-shaped structure about 10 by 15 feet in size. This was covered with closely-woven mats of cattail tied on in one or two layers over the framework. On the very top sheets of elm bark were laid for protection against the heaviest rains.

Simple as they were, the Helton houses provided the residents with a comfortable retreat from rain, wind, and snow. They gave people a measure of privacy. Most important of all, it meant that the Helton people could live for years in one place and grow old in comfort.

Since it was still warm on that fall day, the open ends of the houses were covered with thatched mats. Within a week or two, as the weather turned cooler, the housewives would take these down and, in their place, would hang coverings of deerskin to keep the interiors warm. The hangings served in place of walls at either end of the houses and as "doors," easily pushed aside for people to come in or go out.

Fairly early on this bright fall day, many of the adults began to separate into work groups, each making preparations to pursue a specific task.

Tasks among the villagers were parceled out by group consent. Some were performed by individuals, some by members of an extended family working jointly. When certain jobs, such as harvesting fish, required large numbers of people working together, the work group was composed of people from several clans.

The Helton people organized themselves socially, along rather egalitarian lines (if they operated as the hunter-gatherer societies we have actually observed do). There were no strong political leaders; clan leaders inherited their positions by age. Some leaders were chosen for the purpose of organizing and leading people in the performance of certain very important tasks, such as deer hunting. The men in each clan who had shown the greatest prowess at deer-hunting were the hunt leaders. But this was not always a permanent position. If a hunt leader planned a few hunts which turned out to be unsuccessful, he would be ignored from then on, and a new hunt leader would replace him.

The previous night in the Helton village the hunt leaders from the different clans had met to discuss strategy for the next day's work. They had listened carefully to the counsel of the shaman, who reported to them that the spirit of the deer had indicated that this was a good time to start the annual fall hunting.

The shaman was an interesting person. He had achieved his position after having undergone a religious experience in which he had been "possessed" by a spirit. He played an important role in the community, for it was part of his job to commune with spirits from other worlds and to relay their messages back to the villagers.

To play his role effectively, the shaman had to be a shrewd judge of people, talented at winning them over to his point of view. He also had to have great ability at reading the signs in nature, so that the advice he gave on the proper time for hunting or collecting was correct.

That week, the shaman had gone off alone into the woods for a few days to commune with the spirits. On his return, he had reported to the hunt leaders.

"The spirit of the deer says now is the time for a hunt."

The clan leaders, the hunt leaders, the shaman—in fact, everyone in the village—had been watching the changing weather as the hot torpor of late summer merged into the pleasant, mild days of fall. They were trying to determine the precise time for each of the important hunting, fishing, and collecting tasks that must be done to take advantage of different food resources as they became available and to prepare for the winter.

Soon after the duck-hunters had returned, some of the women left to gather seeds of plants in the river floodplains. They took the children with them, some mothers carrying infants in cradleboards on their backs.

At this time of year the women collected seeds of smartweed (*Polygonum*) and pigweed. We consider these plants common weeds; actually the seeds, which grow in clusters at the apex of the plants, like those of barley, wheat, or oats, are both delicious and nutritious. They also collected the seeds of wild sunflower (*Helianthus*), marsh elder (*Iva*), and goosefoot (*Chenopodium*).

All these plants tend to grow in dense stands wherever stable vegetational cover is disturbed by some force, either natural or human. When the backwater lakes dried up in the summer in Lowilva, these plants were the first to start growing along the exposed muddy shorelines, hence they are called "pioneer." The seeds of these pioneer plants ripen in late October or early November in Lowilva, and the Helton women had to be aware of exactly the right time to harvest the "crops," since the winds might come and within a few days the seeds would be blown to the ground.

The women, working together in family groups, walked through the fields carrying shallow baskets. Some worked alone, others in pairs; they would shake the stalks vigorously to knock the seeds into the baskets. From time to time they emptied the shallow baskets into larger ones. When the large baskets were full, a few women hoisted these on their backs. For ease in carrying, each woman used a tump line, a band of rope or cloth attached to the basket, which the carrier placed across her forehead to help distribute the weight. The party started for home.

Meanwhile, back in the village, as the sun reached its zenith, another groups of women decided it would be a good time to go on a fish-collecting trip since the sun by now had warmed the day a bit. Actually they already had harvested many fish in late summer, since that was the best time for that activity. But on this fall day there was still an opportunity to collect more for their winter stores. The backwater lakes were continuing to dry up, and the water would be shallower, which would make their work easier.

On the way to the lakes, about a mile from their homes, the women talked about which method they would use to collect fish that day. Since there was disagreement among the groups, they agreed to split up, and each party would pursue its preferred method.

At a small lake a group of women, having brought along ground hickory-nut husks, tossed these into the water. The chemicals in the husks, released in the water, poisoned the fish and caused them to float to the surface. As the fish bodies appeared, the women scooped them up in their baskets.

In the other party, a line of women formed and, walking together through the shallow water, stirred up the mud in the lake's bottom. This caused the water to lose oxygen and stunned the fish. The moving line of women herded the sluggish fish to the other end of the lake, into the shallowest water, and their co-workers scooped the fish up into baskets.

There was still a third technique, which they used on occasion. They would build a shallow canal at one of the lakes to drain it of water. This would leave the fish stranded and easy to collect. After scooping up the fish in shallow baskets, the women dumped them into larger baskets. Some of the workers stood over the baskets, pulling out the largest fish, which they filleted. They chose only species of nonfatty fish, such as northern pike, for filleting, since oily fish do not dry as well.

The women of the village had collected their major nut crops several weeks before, when hickory nuts were ripe. However, two women had noticed a nut grove on the north slope of a hillside up in the secondary valley east of the village, where some hickory nuts appeared to ripen later in the season. They walked out to the grove of nut trees, hoping to collect a nice harvest. But to their chagrin, they were able to gather only about a third of a basketful. The squirrels had beaten them to the ripe nuts.

Shortly before the women had left to harvest fish, the deer-hunting parties had returned. The men were pleased; the hunt had gone well.

The hunters had left on their mission well before dawn, because they knew that deer move at very specific times of the day, just before sunrise and again just before sunset. By custom, if they had not caught a deer by an hour after sunup, they would have returned home and gone out again late in the afternoon.

They had gone a few miles away in the upland forests. There they had taken up places along the most traveled deer trails. They had stationed themselves along the trails, each man on a concealed platform in a tree. As a deer came within range, the hunter would hurl his spear at the moving target; the atlatl from which he launched the spear enabled him to double its force for more deadly penetration and greater distance.

Each of the parties had killed a deer. Then the hunter who had made the kill gutted his quarry in the field with help from his companions.

Upon arriving back home, the hunters divided up the meat. This was done according to strict rules based on family relationships. All who shared in the hunt shared in the catch. The men who had killed the deer took the poorest cuts of meat (just as we pass the best food to guests, before we help ourselves). In turn, they would be offered the choicest cuts when others were more successful in bagging a deer.

The men were hungry, and the women stirred up the food in the ever-bubbling "pots" and served them some stew.

After the meal, the hunters relaxed. Some napped. Others sat about gossiping or telling stories, and the afternoon passed pleasantly.

Later, during the winter, the deer-hunters would hunt in small groups. In winter, particularly if there is heavy snow, the deer gather in small bands and move farther up into the secondary valleys, seeking shelter. Working together, the hunters could surprise and kill several deer in a few minutes.

As the men lolled about after their meal, a young woman in her last months of pregnancy, who had remained in the village while the other women went out collecting, took the family "pot" and added some small fish and duck potatoes (roots of the arrowhead plant, *Sagittaria*) to it. She also added some newly hot stones from the fire hearth to the stew, and a bit of water.

Some of the women took fresh deer meat, cut it into thin strips, and spread these out on rocks to dry in the sun. After the meat had been jerked, it would be added to that already in storage for the winter food supply.

Other women spread out the fresh deerskins and began scraping off the fat with stone scrapers. After that, they would hang the skins to dry, and eventually, using bone awls for punching holes, and with bone needles and thread from animal sinews, they would make clothing, coverings for their sleeping benches, and bags from the skins.

By midafternoon most of the women's working parties had returned, carrying their baskets of fish, nuts, and seeds. Some of the women took the larger fish and added them to those already in place in the small, mat-covered smokehouse.

After they had performed these chores and had a bite to eat, the women turned their attentions to preparations for the big feast which was to be held that night.

The fall had been a bountiful one for the Helton people. They were almost tired of all the ducks, geese, and

delicious persimmons on which they had gorged themselves all season. There was no way for them to preserve
any of these foods, so they ate their fill while they lasted.
They had put away huge harvests of dried fish, jerked deer
meat, nuts, and seeds. Because of the bounty of the harvest, the clan leaders, after conferring with the shaman,
had agreed that there should be thanks offered to the spirits of the deer, the river and lakes, and the earth.

The women started to cook in preparation for the feast.
They skewered ducks and geese on a long pole and placed
it over a large roasting pit to cook slowly. Deer haunches
were roasted in a similar manner.

A group of young girls came into the village, bearing
baskets of freshwater mussels which they had collected in
the river's shoals using their bare feet to pick out the mussels among the rocks. They also brought duck potatoes
(the roots of the arrowhead plant, Sagittaria, about the
size of a hen's egg, with a little green shoot growing out of
the side, which resembles the shoots on a bird-of-paradise
flower). These they had dug from the mud banks of the Illinois River, using mussel shells as scoops. They also
gathered ground nuts (Apios americana, sometimes known
as "rosary roots," because they grow on long strings).
These also grow just below the ground surface in muddy
areas.

The cooks placed some of the duck potatoes and ground
nuts into the stew; they wrapped some with leaves and
placed them among the coals to roast. At the last moment
they placed the mussels on the fire to steam.

As the good smells from the cook-fires wafted through
the air, the village came to life, and everyone began to
prepare for the evening's festivities. Women and children
donned special clothing.

The hunters repaired to a corner of the village to paint
their bodies with red pigment made from hematite. The
red powder was used only on special occasions such as
this, when they would dance to honor the spirits, since it
had to be imported all the way from Missouri sources,
fifty miles to the west of Koster.

By now everyone was ravenous, and they gathered near
the fires, getting in the cooks' way, talking and laughing.
The food was ready, but they must first wait for the
shaman to give the opening prayer of thanks.

Finally the shaman emerged from his house. The people

fell silent and made way for him as he slowly, solemnly made his way to the center of the open square. He was wearing a deer-skull headdress, capped with deer antlers, and a deer hide on his back, to simulate a real deer.

The shaman stopped before the clan leaders, turned and faced them, and waited a few moments for absolute silence. Then he began to chant, offering prayers of thanks. He thanked, in turn, the spirits of the deer, the spirits of the rivers and lakes, and the spirit of the earth, for having showered their generosity on the Helton people.

When his prayer was over, the shaman removed his headdress and joined the others. This was the signal for the festivities to begin.

The feast was sumptuous. Besides the roast deer, ducks, and geese, there were roast squirrel, baked fish, and steamed mussels. The stew that night included fresh meat, fish, meat from hazelnuts, redbud pods, and duck potatoes.

There was a porridge made with two kinds of flour, one from acorns, the other from cattail shoots. There were also persimmons, pawpaws, hickory-nut broth, pecans, and walnut meats. And to wash it all down, there was hot tea made from sassafras roots.

The women took care to see that the older members of their family had their share as they sat on the fringes of the milling, laughing crowd. Finally, when everyone had eaten his fill, the shaman stood up and resumed his deer headdress. He picked up a stick, which had been leaning against the wall of his house. A small pillow-shaped stone, called a bannerstone, was hafted at the end of the stick. It was made of orange quartz and beautifully shaped and polished; it symbolized his special role and high rank in the community.

As he walked slowly to the center of the square, the villagers formed a ragged circle, leaving space in the middle for the dancers. The drummers banged down hard to start the music and simultaneously burst into song. As the strong male voices and the voices of the drums rose in the clear night air, the shaman sprang into action. His first dance was dedicated to the spirit of the deer, and after he had been dancing for a while, the hunters joined him, moving vigorously about the floor in imitation of the deer's walk and the hunters' crouch and throw of the spear.

Faster and faster went the drums and the voices. The

dancers leaped in joy, celebrating their successful hunt. Soon, the women, too, began to dance, but they moved more sedately. Little boys imitated the hunters, jumping up and down; the girls took short mincing steps, emulating their mothers.

The clan leaders, seniors in the village, sat together, talking, watching the dancers. They wore their most special outfits in honor of the occasion. One, a tall, straight man, with silvery hair, wore several necklaces. One of these was made of tiny copper beads; another had been strung with the interior coils of conch shells. Like the shaman, he too possessed a bannerstone made of orange quartz hafted at the end of a long pole. It was purely symbolic, reflecting his status in the community. At his belt he wore a knife made of Dongola chert. All of these adornments were made from materials imported into the region from far-off places; their possession signified that he was a man of accomplishment and wealth. In his younger days he had been a very successful hunt leader. Now, as a senior man in his clan, he guilded his people, passing on to them the wisdom he had gained in years of experience.

As the evening wore on, there were other, special dances, dedicated to the different spirits who had helped to make life so good for the villagers.

Near the fire, the old people, wrapped in deerskins to ward off the evening chill, dozed after their full meal.

As I sat there, I could almost smell the delicious aroma from the roast meats; I could see the excited, flushed faces of the dancers; I could hear the beat of the drums and the strong, pulsating sound of the singers' voices, the shuffle of dancing feet, and the snatches of conversation among the watching crowd.

But a real noise from below brought me out of my trance. Because of the warm weather, we had kept the site open a bit later than usual, and a handful of excavators were working. They had apparently found something unusual, for they were clustered about, examining the ground.

I watched them, reflecting that they are fortunate, because they are young and enthusiastic at a time when their chosen profession is becoming capable of revealing much more about the buried past than it ever could before.

It was time to go. Soon, it would be time to close down

the Koster site for another winter and to resume any responsibilities on campus.

But I lingered a bit in the waning afternoon sun. I wished Father Steck were still alive so that I could share our new knowledge of the Amerindians with him. He would have been as pleased as I am that we have been able to retrieve this small segment of the American past and to establish the Koster people as part of our heritage.

Appendix

Changes Needed in Archæology

The new archaeology calls for changes in the way archaeologists practice their profession.

I think that in the future archaeology should be supported by institutions devoted solely to the support of experimentation in archaeological research and education.

Such institutions could build facilities and bring together teams of scientists from different disciplines. They could offer clinical training for future archaeologists. So far, we have turned to the biological and physical sciences for help in practicing the new archaeology. Now we must train our own specialists who can combine expertise in archaeology with in-depth knowledge in one of the biological or physical sciences, engineering, or some other technology. An institute could support experimentation not only in archaeology but also in applied mathematics, civil engineering, computer technology, and other fields, as applied to archaeological research.

Currently, archaeologists must depend on funds from a university or museum, and traditionally the discipline gets a small share of the total funding. An archaeological institute would concentrate on fund-raising specifically for research and experimentation in archaeology.

Some disciplines already have established institutions where groups of scientists work together. The Argonne National Laboratory in Lemont, Illinois, operated by the University of Chicago, carries out broad programs of fundamental research in the physical, biomedical, and environmental sciences and serves as a major center for energy research and development. It also plays a role in the nation's liquid-metal fast-breeder reactor program. NASA, which runs the space program for the United States, is a prime example of the kind of institution I am talking about, except that I envision archaeological institutions as being in the private sector, not under government supervision. In the space program, one brilliant scientist working alone could not have put humans into space nor on the moon, but a great many scientists and technologists co-operating were able to perform spectacular achievements.

At NAP we have taken the first steps toward the establishment of such an institute. We call it the Center for Archaeological Research, and under its aegis we operate the Kampsville Archaeological Center, which is open three quarters of the year—for the spring, summer, and fall quarters in the Northwestern University academic year.

NAP operates ten laboratories and a computer program at the Kampsville Archaeological Center, several of which are open on a year-round basis.

In addition, the Kampsville Archaeological Center was conceived and is being operated as the archaeological equivalent of a medical center where future physicians train. Our students, in addition to learning excavation techniques, rotate through various laboratories, learning about different kinds of data analysis, just as internes and residents rotate throught medical clinics.

Eventually we hope to provide clinical training at an archaeological institute on the Northwestern campus in Evanston too. We chose to develop the Kampsville Archaeological Center as a clinical training program first, not only because of its proximity to sites but also because real estate there is much less expensive than in Evanston.

The Center for Archaeological Research sponsors field schools for students from junior high school through graduate school levels, and adult field schools, as well as various academic conferences throughout the year.

—S.S.

Bibliography

Asch, Nancy B., Richard I. Ford, and David L. Asch. *Paleoethnobotany of the Koster Site.* Illinois State Museum Reports of Investigations, no. 24 Springfield, Ill., 1972.

Baldwin, Gordon C. *The Riddle of the Past.* New York: W. W. Norton & Co., Inc., 1965.

Brown, J. A., S. Struever, D. L. Asch, N. B. Asch, C. A. Bebrich, J. E. Buikstra, T. G. Cook, F. C. Hill, M. E. W. Jaehnig, J. Schoenwetter, and R. K. Vierra. "Preliminary Contributions of Koster Site Research to Paleoenvironmental Studies of the Central Mississippi Valley," to appear in *Approaches to the Study of Man and Environment in New World Archaeology.* To be published as a Memoir of the Society for American Archaeology.

Buikstra, Jane E. "Cultural and Biological Variability: A Comparison of Models." Paper presented at the annual meeting of the American Association of Physical Anthropologists, 1974.

Butzer, Karl W. *Geomorphology of the Lower Illinois Valley as a Spatial-Temporal Context for the Koster Archaic Site.* Illinois State Museum Reports of Investigations, no. 34. Springfield, Ill., 1977.

Ceram, C. W. *The First American.* New York: Harcourt Brace Jovanovich, Inc., 1971.

Clark, Grahame. *Prehistoric Europe: The Economic Basis.* London: Methuen, 1952.

Cook, Thomas Genn. "Tools and Tasks at the Koster Site, West-Central Illinois." Paper presented before the annual meeting of the Society for American Archaeology, 1974.

Deuel, Leo. *Conquistadores Without Swords*. New York: St. Martin's Press, Inc., 1967.

Fowler, Melvin L. *Cahokia: Ancient Capital of the Midwest*. Addison-Wesley Module in Anthropology, no. 48. Reading, Mass.: Addison-Wesley Publishing Co., Inc., 1974.

Haag, William G. "The Bering Strait Land Bridge," in *Readings from Scientific American*. San Francisco: W. H. Freeman & Co., 1973.

Haynes, C. Vance, Jr. "Elephant-Hunting in North America," in *Readings from Scientific American*. San Francisco: W. H. Freeman Co., 1973.

Hill, Frederick C. "Effects of the Environment on Animal Exploitation by Archaic Inhabitants of the Koster Site, Illinois." Ph.D. thesis, University of Louisville, Louisville, Ky., 1975. Ann Arbor, Mich.: Xerox University Microfilms.

Hole, Frank, and Robert F. Heizer. *An Introduction to Prehistoric Archaeology*. New York: Holt, Rinehart and Winston, Inc., 1973.

MacNeish, Richard S. "Early Man in the Andes," in *Readings from Scientific American*. San Francisco: W. H. Freeman & Co., Inc., 1973.

Samachson, Dorothy, and Joseph Samachson. *Good Digging*. New York: Rand McNally & Co., 1970.

Schoenwetter, James. *The Koster Site: A Progress Report*. Scientific Papers Series, Northwestern Archaeological Program. In press.

Stuart, George E., and Gene S. Stuart. *Discovering Man's Past in the Americas*. Washington, D.C.: National Geographic Society, 1969.

Taylor, Walter W. *A Study of Archaeology*. Carbondale and Edwardsville, Ill.: Southern Illinois University Press, 1973.

Watson, Patty Jo, Steven A. LeBlanc, and Charles L. Redman. *Explanation in Archaeology*. New York: Columbia University Press, 1971.

Index

Illinois State Geological Survey (ISGS), 41, 42, 43, 101
Intermarriage, 184, 198

Jaehnig, Jan, 113
Jaehnig, Manfred, 28, 111
Jersey Bluff people, 9–10, 11, 13, 14, 16, 18, 19–20, 114, 135, 139, 147, 173, 216, 217, 220
Juglone (hormone), 132

Kamp, Aloysius (Pete), 45–46
Kamp, Joe, 45
Kamp, M. A., 27, 45
Kamp Mound Number 9, 46, 47, 50, 57, 94–95, 173
Kampsville Archaeological Center, 241
Kampsville headquarters, 25–28, 31–32, 36, 37, 45, 49–53, 57–58, 62, 66, 72, 92, 102, 105, 111, 143, 146, 163, 166, 196
Kapka, Robert, 37, 210
Kelly, A. R., 50
Kennedy, John F., 171
Klein, Richard, 48
Klunk mounds, 15
Koski, Ann, 164
Koster, Mary, 69–73, 147
Koster, Theodore (Teed), 3, 5, 11, 13, 15, 16, 27, 30, 31, 64, 67, 69–73, 77, 111, 114, 147, 225
Koster site: climate, 91–104; controlled survey of, 15–17; daily life (3500 B.C.), 225–39; dirt removal, 33–34, 61–63, 99; earliest datable proof of people, 40–41; first digs, 18–24; fund raising for, 64–68; importance of, 23–24; invader theory and, 184–99; location of, 3–4, 8–9
!Kung San people, 158,

221–22
Kupcinet, Irving, 66

Leakey, Louis, Mary, and Richard, 184–85
Lee, Richard, 222
Libby, Willard F., 38
Loess soil, 17–19, 23
Longacre, William, 48
Lower Illinois River Valley, 42–57; archaeological sites, 51–52; environmental events, 105–16; population (A.D. 800), 11; settlement of, 42–43. See also Hopewell culture; Koster site
Lyell, Sir Charles, 79–80
Lynch, Thomas, 48

McAdams, William, 220n.
McJunkin, George, 40
MacNeish, Richard S., 40, 84
Macoupin site, 8, 14, 124–25, 126, 128
Maintenance technology, 159
Malthus, Thomas, 132, 133
"Markers" on bones, 106
Marsh elder, 89, 118, 122–23, 227, 232
Martin, Paul, 167
Maze, W. H., 51
Metate, 138–39, 209
Meyers, Thomas, 162, 163
Midden, 18, 19
Mississippian period, 43
Mississippians, 176, 193, 198–99, 216, 217, 220–21
Modoc Rock Shelter, 137, 194–95, 196
Monks' Mound, 220
Morgan, Lewis Henry, 81
Moss, Donald, 67
Mussel shells, 149; ratio of strontium to calcium, 110; reading past environment from, 105–16
Muto, Guy, 163, 164

Recommended Reading from PLUME and MERIDIAN